Authors, Users, and Pirates

The Information Society Series
Laura DeNardis and Michael Zimmer, Series Editors

Authors, Users, and Pirates

Copyright Law and Subjectivity

James Meese

The MIT Press
Cambridge, Massachusetts
London, England

This book was set in Stone Serif by Westchester Publishing Services.

Library of Congress Cataloging-in-Publication Data

Names: Meese, James, (Writer on law), author.
Title: Authors, users, and pirates : copyright law and subjectivity / James Meese.
Description: Cambridge, MA : The MIT Press, 2018. | Series: The information society series | Includes bibliographical references and index.
Identifiers: LCCN 2017023974 | ISBN 9780262037440 (hardcover : alk. paper) ISBN 9780262549653 (paperback)
Subjects: LCSH: Copyright. | Copyright--History. | Authorship. | Copyright infringement.
Classification: LCC K1420.5 .M44 2018 | DDC 346.04/82--dc23 LC record available at https://lccn.loc.gov/2017023974

For Adrian and Benjamin

Contents

Acknowledgments

The ideas that sit at the heart of this book emerged from my dissertation. They only came to fruition thanks to an incredibly supportive supervision team consisting of Esther Milne, Ramon Lobato, and Jock Given. I thank them for the invaluable mentorship they provided throughout my doctoral studies.

Thanks to Isabella Alexander, Christopher Comerford, Angela Daly, Liz Giuffre, and Katrina Schlunke for suggestions and comments on chapter drafts. I am particularly grateful to Kathy Bowrey, who read numerous drafts and regularly challenged me to produce better work; Alan McKee, who generously read an early draft of this book; and Tim Laurie, who posed a series of productive conceptual questions that helped me refine my argument toward the end of the writing process. I also thank Eva Hemmungs Wirtén, who invited me to contribute to a range of collaborative interdisciplinary endeavors that allowed me to think more broadly about the role and function of copyright. I also extend my gratitude to the anonymous reviewers of the proposal and manuscript. They all offered generous and considered feedback, which was critical in shaping the approach and direction of this monograph.

I am lucky to work among a vibrant and inspiring scholarly community broadly focused on the study of intellectual property, media, and culture. Instead of offering a list of names, I simply want to express my gratitude to all of my colleagues for the advice and friendship they have offered along the way. I am particularly indebted to my colleagues at the University of Technology Sydney (UTS) and at my former institutions—Swinburne University of Technology and the University of Melbourne—for their collegiality and support. I also thank the excellent team at the UTS Faculty of Arts and Social Sciences Research Office, Wes Mountain for his excellent diagrams, and Nick Jarvis for his assistance with some last minute corrections. I am especially grateful to all the students I have taught at UTS for their willingness to discuss ideas, challenge my thinking, update my knowledge, and make me laugh.

Thanks to Michael Zimmer and Laura DeNardis for their initial enthusiasm, Virginia Crossman for her careful editing, and Gita Devi Manaktala, Emily Taber, Jesús Hernández, and Susan Buckley at MIT Press for all of their assistance and advice.

For their love, encouragement, and laughter, I thank my parents, my two awesome "sibz," and all of my friends. Finally, I want to thank my partner Caitlin for her incessant interest in this project, for offering (and sometimes demanding) to read a chapter draft, and for keeping up the fight when so many would have given up. You're the best.

I acknowledge the Faculty of Arts and Social Sciences at the University of Technology Sydney, who generously provided financial support that assisted in the preparation of this manuscript.

I also wish to acknowledge SAGE and Litwin Books for allowing some previously published material to be incorporated into this book. Some parts of this book draw on the following articles:

Meese, James. "The Pirate Imaginary and the Potential of the Authorial Pirate," in *Piracy: Leakages from Modernity*, edited by Martin Fredriksson and James Avanitakis, 19–37. Sacramento, CA: Litwin Books, 2014.

Meese, James. "User Production and Law Reform: A Socio-Legal Critique of User Creativity." *Media, Culture & Society* 37, no. 5 (2015): 753–767.

Introduction: Copyright Law and Subjectivity— A Relational Approach

Copyright law is a topic of significant public interest. No longer limited to legislators, entertainment industries, and a handful of academics,[1] the phenomenon of pervasive digital connectivity has made copyright a first-order political and social issue in the four jurisdictions this book focuses on— Australia, Canada, the United Kingdom, and the United States. In these countries, people have engaged in numerous activities that may be classed as copyright infringement, such as downloading movies and music from torrent sites or sharing music with friends (and strangers) through cyber lockers or remote servers located in the cloud. Homes have also become centers of creative production with individuals uploading works that often draw on existing copyrighted material and making them available to (potentially) massive audiences.[2] The prevalence of these activities means that copyright law is implicated in the routines and habits of ordinary people in a more direct fashion than ever before and, subsequently, people have become increasingly aware of copyright law.[3] Although the average person will not know what section 111 of the Copyright Law of the United States refers to (I picked a section at random, but for those interested, section 111 addresses limitations on exclusive rights: secondary transmissions of broadcast programming by cable), they will most likely have an opinion on the rights and wrongs of online piracy and a general stance on what issues copyright law should focus on.

Institutional actors in the copyright sector have noticed this transformation in domestic media consumption as well as the emergence of amateur media production, and responded in a variety of ways. For example, industry lobby groups (typified by the vocal and well-resourced Motion Picture Association of America or MPAA) have decried online piracy. These groups have demanded that governments reform copyright law to better protect

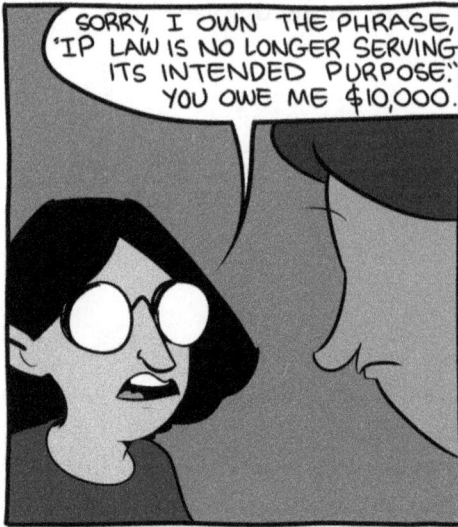

Figure 0.1
A broader conversation around the role and function of intellectual property has been occurring for some time now. Credit: Zach Weinersmith, smbc-comics.com /comic/2014-05-29.

particular creative sectors, engaged in sustained educational campaigns with the aim of changing consumer practices,[4] and begun court proceedings in several jurisdictions.[5] In response, a number of pirate parties have emerged across the world, challenging these attempts to strictly enforce copyright law.[6] Activist groups like the Electronic Frontier Foundation have also been vocal critics of industry groups (and their practices), as well as a strong public voice for user rights. Alongside this fervent debate, policymakers and think tanks have examined the growth in amateur creativity and considered what it means for the creative industries.[7] In short, an ongoing cultural conversation about copyright law is occurring both in people's everyday life and at the more rarefied level of public policy and law (see figure 0.1).

There has been one notable phenomenon throughout these discussions. The author, user, and pirate have each retained a constant presence in these debates. Advocates for particular creative industries regularly invoke authors as they call for more legal protection. Users are also prominent, with user rights and user-generated content commonly discussed. The pirate is another conspicuous subject, with complaints about online piracy and the emergence of pirate parties placing this subject at the heart of discussions about the current operation and potential future of copyright law. Indeed, although other actors are present in these discussions (such as various intermediaries, from publishers to search engines), as we will see throughout this book, they often strategically take up the mantle of an author or user in policy debates.

Of course, these subjects have not appeared out of nowhere.[8] The author and user have been prominent subjects in copyright law for some time. For example, although there are differences between jurisdictions, it is not a controversial statement to note that copyright law is particularly interested in authors and has a lot to say about them. According to copyright law, if you create an original work and it is embedded in a material form, you are an author. Legislation outlines the rights you hold and the exceptions and limits to them in some detail. Many civil and common law countries also have moral rights, a system of rights based around the personal identity of the author, and so the author is fleshed out even further in law.[9] The user is also present in law, albeit to a lesser degree. The United States has fair use and other common law jurisdictions have fair dealing exceptions, which allow prospective users to use copyrighted works in particular situations (fair dealing)[10] or with reference to particular guidelines (fair use).[11] In addition to this, there are a number of limits to the authorial remit that are clearly introduced with users (or prospective authors) in mind, such as the idea/expression dichotomy (the principle that a person can copyright their individual expression but not an idea or fact), or the ability to sell a copyrighted work once you have purchased it. The pirate is not named in legal texts but "infringement" is a common term in the law, which immediately raises the specter of the pirate. Moreover, Adrian Johns has also shown that the pirate retains a strong presence in the cultural history of copyright law, and has played a significant role in how people have thought about copyright since its earliest days.[12]

So, at this point we can identify three central subjects who structure our overall experience and understanding of copyright law: the author, the user, and the pirate. These subjects are present in conversations around the law, as well as within the law itself; they give us a range of subject positions to inhabit and in turn offer us a set of agencies. We can "author" an original work, "use" an existing copyrighted work, or "pirate" it by engaging in copyright infringement. The presence of these subjects becomes more notable once we consider the fact that laws are reformed, new precedents are occasionally set in common law, and the broader technological and cultural context in which copyright operates regularly changes. So, with these legal processes and divergent contexts in mind, some questions start to emerge: How did this tripartite constellation form through this process of jurisprudence and attendant social and technological development, and why are these three subjects so resonant today?

At the moment, we do not really know the answer because there is a lack of clarity around the socio-legal construction of subjectivity in copyright law. Indeed, various authors have argued that the author,[13] user,[14] and pirate[15] are conceptually underdeveloped. *Authors, Users, and Pirates* responds to this situation by offering a detailed account of how these subjects are constituted in copyright law and, in doing so, takes a step toward a fuller conceptualization of all three subjects. As part of this account, the book argues that subjectivity in copyright law should be approached relationally and, to that end, offers a relational framework that encompasses the author, user, and pirate. Departing from existing modes of legal scholarship that view authors and users as the dominant audiences of copyright law,[16] this approach views these subjects as interconnected, a position that is informed by relational approaches already available in legal scholarship[17] and understandings of subjectivity that have long been prominent in cultural theory.[18] The framework seeks to bring these subjects together to offer a new perspective on the formation and development of subjects in copyright law, challenging accounts that structurally and symbolically disaggregate the author, user, and pirate. The approach has a historical lineage but is also motivated by recent changes in the production, distribution, and consumption of media, which raise further questions about these subjects. With online copyright infringement supporting media consumption for much of the late twentieth and early twenty-first centuries, can we still view pirates as marginal figures of consumption wholly separate from users and authors? Should the creative acts that people regularly engage in online still be called "user generated"? These questions shed light on the cultural and legal ideologies that attempt to keep these subjects formally separated, and subsequently highlight the fact that these subjects may not be as distinct as first thought.

In short, *Authors, Users, and Pirates* will make the case that we need to approach subjectivity relationally and attend to all three subjects if we are to understand how they are constituted and sustained over time. Addressing the relationships between authors, users, and pirates will clarify how the key conceptual underpinnings of copyright law are justified in practice through doctrine and jurisprudence as well as highlight the critical role of culture and everyday practices in shaping and defining these three subjects. We will explore the codependence of these subjects and the different ways these relationships are organized over time through a range of contemporary and historical examples in the chapters that follow. Through this process, we can attend to moments of tension around how these subjects are constituted and defined in relation to one another, which often emerge only during points of contested cultural and legal interpretation. This, in turn, allows *Authors, Users, and Pirates* to address a number of key questions central to the future of copyright law: How do key actors think the audience of copyright law should be structured? How are these three subjects co-constituted by legal institutions, broader cultural understandings of copyright law, intermediaries, and individuals? What identity work is engaged in to make sense of the complicated processes that support creation? And, as we look ahead, what does this account of relational subjectivity tell us about the role and function of copyright more generally?

Studying and Situating the Author, the User, and the Pirate

Before outlining the proposed relational framework of subjectivity, it is worthwhile to briefly outline the existing scholarly conversation around these subjects. Copyright law is often positioned as a law for authors,[19] so it should come as no surprise that a significant amount of attention has been paid to the authorial subject. Still, it is of interest that the most notable early analyses of the author emerged from interdisciplinary scholars—Martha Woodmansee,[20] Peter Jaszi,[21] and Mark Rose[22]—who were influenced by the work of Michel Foucault[23] and Roland Barthes.[24] These researchers were committed to deconstructing the author, challenging its uniqueness as a legal subject, and placing copyright law in a broader historical and theoretical context. Their central argument was that a bias toward the "romantic" concept of an authorial genius had always been present in copyright law's understanding of the author. In the following years, legal scholars have taken up this interdisciplinary theorization of the author and in doing so have offered useful corrections to this history, focused namely around the bias toward romanticism. It has been noted that authorship is indeterminate[25] and that the development of legal doctrine has not always mapped onto this ideological position.[26] Subsequently, a dominant narrative about creation has not been sustained in law across time. Indeed, as we will see in chapter 1, the question of who can be an author and what counts as authorship under copyright law has troubled the minds of judges, lawyers, and scholars across the nineteenth and twentieth centuries. This has left the author and the notion of authorship as at best a bundle of competing ideologies[27] and at worst an "intellectually unstable [and] essentially bankrupt"[28] category, underdeveloped both conceptually and doctrinally.[29]

As well as offering a series of useful corrections to the germinal interdisciplinary works that sparked these conversations around authorship and the author, this later body of legal scholarship already tells us something about subjectivity in law and how the author relates to the user and the pirate. These critiques show us that rather than a dominant socio-legal narrative of romantic genius defining the authorial subject, the constitution and positioning of authorial subjectivity in culture and law has changed over time, gesturing to the malleable nature of copyright's subjects. The recognition of the author as an indeterminate actor reiterates this point, suggesting that there are moments where an "author" may not be accurately discerned and may alternatively be identified as a "non-authorial" user or even a pirate. But despite this recognition of the relational nature of authorship, much of this extant work has not significantly engaged with these alternate subjects. Although pirates, users, or publishers may be mentioned briefly in these texts, they generally feature as background characters while scholars examine the author's changing status across history.[30]

That being said, another group of legal researchers has moved beyond this author-centric model of analysis to account for other subjects of copyright. The work of Lyman Ray Patterson and Stanley Lindberg has advocated for a detailed account of the user, and has recognized that the author and user are connected to one another in copyright law.[31] Indeed, they acknowledge some sort of relationality in copyright, stating that "to define one person's rights in relation to a work is to define the rights of all other persons relevant to the work."[32] In more recent years, Joseph Liu has offered a useful theorization of the consumer,[33] Johanna Gibson has examined the different ways the author and user can influence each other,[34] and Alina Ng has addressed the phenomenon of users authoring works online and becoming authors.[35] These scholars largely focus on the interrelations between author and users, which provides a useful account of how the author and user are relationally constituted in formal law but does not offer much insight into how broader cultural phenomena inform the constitution of subjectivity in law. This focus also does not recognize that the act of infringement is linked with the pirate, a subject that operates in relation to the author and user and (as we will see) informs socio-legal understandings and interpretations of copyright. Therefore, although these analyses can tell us about the formal legal structure of copyright, they can tell us less about the cultural imaginaries of copyright and how they intersect with legal processes.

Conversely, the pirate has become a topic of interest for many cultural studies scholars. Patrick Burkart and Martin Fredriksson[36] offer nuanced critiques of piracy exploring the "boundary maintenance"[37] performed by this subject and examining the contrasting cultures that emerge from this practice, from mainstream consumption to political engagement. Other scholars contend that copying has a productive capacity, one that is particularly important in establishing infrastructures of distribution in the Global South.[38] These are important observations that will go on to inform my own advocacy for incorporating the pirate into a relational framework of copyright law. Still, the subject of the pirate has largely been analyzed somewhat apart from the legal doctrine and jurisprudence that animates the concept of infringement.

A select group of scholars has attempted to explore the relationships between all three subjects of copyright law. US legal scholar Rebecca Tushnet has challenged the broad interpretation of non-transformative copying as a passive activity entirely devoid of cultural weight.[39] She instead argues that when read in relation to the first amendment of the US Constitution (the right to free speech), copying can contribute to the development of self-expression through the shared consumption of cultural material and subsequently can be said to require both judgment and creativity.[40] Tushnet does not comment on subjectivity but, by making this claim, she challenges both dominant socio-legal conceptualizations of the pirate as infringer and accounts of "copying" that locate it as an aberrant act with no redeeming factors.[41] Similarly, Julie Cohen offers an account of the user that anticipates the central concerns of this book by questioning attempts to place firm boundaries between the author, user, and pirate.[42] Her concept of the "situated user" attends to authorship, use, and infringement, with Cohen noting that a user may engage with copyrighted works in a range of ways.[43] She makes the case for a nuanced account of the user in copyright law that allows people to engage with copyrighted works in a flexible manner.

There is already a significant body of research around these three subjects and recognition that these subjects are connected. But there is more work to be done. The analytic focus of the scholars mentioned above tends to coalesce around one or at most two subjects of copyright law. Cohen addresses a range of potential capacities available under copyright law, from infringement to creation, but she narrows the focus of her inquiry to the user.[44] Tushnet acknowledges the complications of law but does not reflect on how her conclusions inform the broader conceptual framing of the subjects under copyright law's remit.[45] Patterson and Lindberg,[46] Gibson,[47] and Ng[48] establish a relationship between author and user but do not address (or leave no space for) the pirate as a separate subject. Conversely, cultural research around piracy tends to rest on a binary opposition between the author and the pirate, and scholars can minimize the importance of jurisprudence in processes of subjectification when conducting these analyses.[49] Although there is some recognition of a relationship between these subjects, there is not yet an overarching conceptual frame that encompasses the author, user, and pirate and allows us to explore the various ways that subjectivity manifests in copyright.

The relational approach offered by this book allows these issues to be addressed by viewing copyright law as a legal structure that continually balances these three subjects. From this conceptual platform, it is possible to conduct an overview of the tensions, conflicts, and other interactions between subjects without privileging one or two as the dominant objects of analysis (although that approach brings its own benefits). As well as providing a detailed account of how each subject relates to the others, this approach provides a useful way to think about the dynamic and indeterminate nature of all subjects under the remit of copyright law. The next section provides more detail about the proposed relational approach to subjectivity and places this framework in conversation with existing scholarship on legal subjectivity, critical theories of subjectivity, and ideology more generally, as well as a growing body of research that promotes a relational approach to law.

Author-User-Pirate: The Relational Triad, Subjectivity, and Law

Author, Users, and Pirates develops the contention that copyright's subjects are relational and begins its analysis from the premise that these three subjects are meaningfully connected to each other. To do this, the book will use the aforementioned relational triad, a conceptual framework visually represented as a Venn diagram (see figure 0.2). The framework acknowledges that each subject retains an independent domain but also recognizes that interconnections occur between subjects, leaves space for interstitial subjects between these fully fleshed out subjects, and provides a way to visualize how they change over time.

I will provide further context around the relational triad and my approach to subjectivity and relationality by articulating how they relate to existing academic scholarship. The discussion begins with a consideration of subjectivity. My aim in this section is not to parse complicated scholarly arguments but rather to outline a viable and useful approach to the subject that will underpin the chapters that follow. Therefore, I will

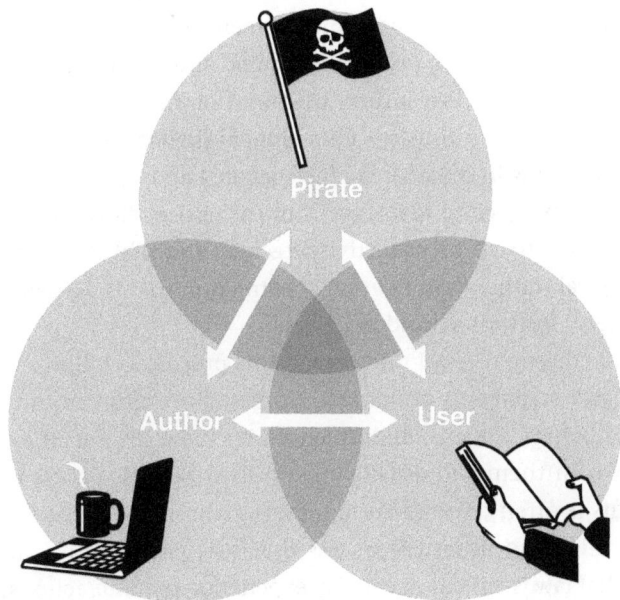

Figure 0.2
The relational triad.

discuss theorists generally rather than exhaustively. I will then move on to relationality and examine how recent legal scholarship has outlined a productive conceptual framework that can be used to examine legal subjectivity. At this point it is also worth noting that the interdisciplinary approach taken here means I will be speaking to both cultural researchers and legal scholars. Therefore, at times this book will discuss law in a fashion that may seem relatively basic to legal scholars and, conversely, I will occasionally dwell on theoretical premises that cultural studies scholars would take for granted. I hope that you find any new discussions from alternate disciplines interesting, and disciplinary histories and debates that are already familiar to you still engaging.

Subjectivity

Authors, Users, and Pirates approaches subjectivity as a social phenomenon and as a form of internal experience but is predominantly interested in the broader social construction of subjectivity. In the chapters that follow, the book outlines how the author, user, and pirate are constructed as legal subjects and examines how individuals come to be interpellated under copyright law. It also addresses the experiential element of subjectivity on occasion and explores how different people understand and construct their own subjectivity in relation to the author, user, and pirate. This overall approach is informed by Louis Althusser's foundational theorization of the subject, which outlines how individuals are interpellated and become subjects, and subsequently "subject to" ideology.[50] Critical legal studies scholarship retains a similar approach to subjectivity, recognizing that the legal subject is an ideological (rather than neutral) construction that is partially created by various legal institutions and practices.[51]

In the context of copyright law, I understand this process as follows.[52] Legal institutions and statements about law are based on ideas around how people should behave. People are influenced by these ideas as they come into contact with them and usually amend their behavior accordingly (however, this is often done in an imperfect fashion). It is important to note that this is an informal process that sees people draw on broad notions of what law is "in our day to day experience and interactions in the world" rather than specific legislation or cases.[53] Therefore, in the context of subjectivity, copyright law is understood as "the set of legal notions we rely upon to bound our activity and to suggest the possibilities of behavior," in addition to the legislation and judicial precedents that make up formal law.[54] Subsequently, legal frameworks do not simply represent the world but rather bring into being particular actions, behaviors, and interactions between people that are formalized through particular subjects.

The Althusserian approach described above is broadly determining, with subjects defined through legal infrastructures and larger social, cultural, and historical trajectories. Yet the emergence and maintenance of subjects and the process of interpellation is not an entirely top-down process. As the preceding discussion notes, people's everyday experience of copyright and conversations about the law contribute to the broader social construction of subjectivity. Indeed, Rosemary Coombe notes that once transformative power is granted to the "everyday struggles over meaning," it becomes possible to recognize that "people's anticipations of law ... may actually shape law."[55] Subsequently, *Authors, Users, and Pirates* recognizes that everyday understandings of authorship, use, and piracy directly assist in the construction of those subjects, although these actions are all ultimately influenced by broader historical, social, and cultural factors.

This recognition of agency within a broader determining ideological structure also allows for a more flexible notion of interpellation to be advanced. Following the work of Coombe and Paul Smith, I suggest that "ideological interpellation is never total or complete."[56] There is a space for the subjective actor to resist their interpellation and even for interpellation to fail. Whereas Althusser presents interpellation as the totality of a person's identity, both Smith and Coombe articulate alternative theories of the subject that allow "for gaps and fissures in the agent's [i.e., individual's] experience of interpellative messages" despite the presence of an available subject category (i.e., "the author" or "the user").[57] Therefore, although institutions and individuals are always interpellated somewhere across the author-user-pirate framework, individuals and institutions can attempt to fashion their own discourse through which to legitimize their activity (i.e., claiming to be users rather than pirates, or authors rather than users). My approach sees ideology directly inform the production of particular subjects, but also affords individuals the capacity to potentially deploy a "contestatory subjectivity" within the context of the relational triad.[58]

In addition, I view the relational triad as a flexible (rather than a fixed) entity. In the chapters that follow we see the relationships between these subjects organized in a variety of contextually and historically specific ways, to the extent that some subjects are minimized in the triad at times only to reemerge later on. This process establishes the relational triad as a constantly changing procedural framework, providing a space from which to explore how the triad is constituted in different social contexts and jurisdictions as well as across different moments in history. Although the relational triad retains a constant presence, it is mobilized in a variety of ways. This allows the triad to function as a useful analytical tool that can examine the varied ways that institutions, industries, and individuals collectively understand and co-construct the subjects of copyright law.

Relationality

The approach to subjectivity just described provides a basis from which to account for relationality. As noted, the relational triad does not just outline a broad framing of subjectification in copyright law but also recognizes that these subjects are connected and maintain relationships with one another, which can reconfigure or change over time. I develop this claim by moving a scholarly conversation around relationality and legal rights away from individual and collective relationships and toward an analysis of relationships in a more abstract sense. Before detailing this departure, I will briefly outline the key concerns of the scholarly literature around relationality and law. This body of work tackles one of the dominant approaches to legal subjects: that they are individual and autonomous beings. This conceptual position is grounded in liberal political theory, which establishes law by outlining a boundary of rights around an individual, providing a barrier that protects the individual "from intrusion by other individuals or the state."[59] Copyright law stands as a classic example of this genre, with the prospective author given a litany of rights and protections to secure their ownership of the work in question. Relational theorists argue, however, that this arrangement fails to recognize that "people are not self-made."[60] As Jennifer Nedelsky points out, "[w]e come into being in a social context that is literally constitutive of us."[61] Our relationships and interactions with each other and with various institutions define who we are.

Subsequently, an individual cannot exist outside of the relationships that structure and define them, since a person's identity is always partly "constituted by ... interactions with others."[62] This assertion destabilizes the foundational assumption of liberal theory—that there is a "unitary self" that exists a priori.[63] A relational analysis carries direct implications on how subjectivity is constituted and understood in law. If a political and legal identity is partly constituted by relationships, then the social and cultural spaces from which these relationships emerge also become points of interest.

The preceding analysis then raises the issue of what replaces the discourse of rights that has typified our legal language for so long. To manage these competing tensions, a relational approach reimagines the very notion of political autonomy. For relational theorists, it is only through relationships that autonomy is made possible: one is not able to be fully autonomous without other people.[64] Acknowledging this concept of situated autonomy forces us to reconfigure our given notions of law and its operation within society. No longer do autonomous individuals, wholly separate from society, possess legal rights. Instead legal rights are viewed as vehicles through which relationships are constructed between actors.[65] This conceptual shift does not deny the possibility of power within these relationships, but again emphasizes the relational and situated aspect of these commitments.[66] Carys Craig has introduced relational legal theory to copyright scholarship and offered the most compelling and consequential deployment of this approach.[67] She explains that because copyright law is predicated on a liberal political theory based on Enlightenment ideals of "individuation, detachment, segmentation, and abstraction," our understanding of this law is shaped by "doctrinal categories" that set up unhelpful binaries of "creation/reproduction, author/user and laborer/free-rider," which my earlier critique of the binary analyses of subjects in copyright law echoes.[68] Craig uses relational theory to challenge copyright law's existing construct of the self as well as offer "a new theoretical model" centered on the "relational self."[69]

Craig's model offers a new conceptualization of the author-subject as a relational author who "is always already situated within, and constituted by, the communities in which she exists, and the texts and discourses by which she is surrounded."[70] This allows us to view the author and their creative impulses as part of a broader "network of social relations" rather than an independent creator, and has direct implications for copyright law.[71] No longer is the author's right reducible to an "individual entitlement," with the author separated from the wider public.[72] Instead copyright law must be understood as a process by which particular relationships are structured between authors and users, "allocating powers and responsibilities amongst members of cultural communities."[73]

The major benefit of Craig's relational approach to copyright law is that it challenges traditional conceptualizations of creativity in law. She suggests that a relational model will allow us to develop a model of authorship that encourages the "derivative, collaborative and communicative nature of creativity."[74] Her attempt to redefine authorship stands at the heart of the relational model in copyright and others have attempted to develop legal principles based on her approach.[75] Still, the broader literature around law and relationality and Craig's own deployment of this model offer a wider spectrum for intervention, beyond debates around authorship, creativity, and the public interest. By challenging such fundamental considerations of copyright and the philosophical basis of subject formation in law, these critiques have a direct impact on how we conceive of subjectivity.

A relational approach provides an avenue through which to focus on abstract relationships between different subjects of copyright law.[76] Although this departs from existing relational approaches to law, which focus productively on rights, relationships, and the "particularity of the individual,"[77] I suggest that turning toward subjectivity provides further insight into how copyright law and the actors that surround it attempt to manage and negotiate relationality. This process moves the analytic lens from "rights" to "roles" and acknowledges the inherent interdependence of the author, user, and pirate. Therefore, although this approach addresses and analyzes complications around rights, it also moves beyond these classic deployments of relationality and uses the concept in a broader sense to examine the categorization and construction of subjects and the interrelations between them. The following section will provide further detail around the benefits of using a relational approach to explore copyright and subjectivity.

The Case for the Relational Triad and the Study of Subjectivity

I have made the case that subjectivity is not sufficiently accounted for in existing studies of copyright and that no one has studied all three subjects of copyright law concurrently. But what are the benefits of engaging in this task? As we will see throughout this book, a relational approach sheds light on a few areas of copyright law. First, this study allows us to examine how subjects are constructed in the everyday operation of copyright law. Currently, the author, user, and pirate are conceptually weak and under-theorized as legal categories and, as noted, some work has been done to bring clarity to these subjects individually.[78] Developing a detailed account of a single subject is particularly valuable in discerning how that subject is constructed across a historical period, but it does not tell us much about how subjectivity in copyright law functions in practice. The relational triad extends these studies by accounting for the relational co-construction of subjectivity that occurs in copyright. This phenomenon can be identified in the slightly varied ways in which different jurisdictions approach the author, user, and pirate, the critical points at which subjects intersect with one another in culture and law, and the bargaining and trading around copyright that occurs in national and transnational forums that implicates these subjects.[79] In short, this relational approach leaves space for the mess-iness and indeterminacy of copyright as a legal framework, policy instru-ment, and cultural force that regularly ends up addressing all three subjects simultaneously.

Second, the relational approach also fits in neatly with copyright's own historical trajectory and its function as a complicated and occasionally con-tradictory legal framework. As I will discuss further in chapter 1, the lack of an originating theory or concept animating copyright means that copy-right has always been relational, with people working out the capacities and agencies of the subjects of this law across the centuries. This has been a legal task (i.e., what rights does an author have?) as well as a cultural task (i.e., who are authors and how do we think about them?). Subsequently, the relational triad stands as a useful way to understand how critical decisions have been made around the scope of authorship, infringement, and use (to give one example) and to account for the socio-legal discourses that have informed these processes.

Finally, studying subjectivity provides a way to explore the various nar-ratives that are constructed for copyright within and beyond law. It shows us how particular legal and cultural actors imagine the beneficiaries and audiences of copyright law and provides a space from which to explore why some modes of subjectification are successful and others fail. In addition, the relational framework also allows us to investigate moments of crisis and confusion in these narratives by excavating particular acts (like authorship or copying) and exploring how they can operate relationally across various subjects.

In short, thinking about copyright relationally allows us to understand the law as it is currently constructed. Considering the conceptual uncertainty underpinning copyright law,[80] it makes sense to attend to all three subjects collectively rather than presuming to place one or another subject at the heart of copyright's concerns. The triad stands as a useful position from which to begin a sustained examination of copyright's past and present and articulate a complex approach to subjectivity that will examine how these three subjects were formed and how they have subsequently been redefined in response to a range of technological innovations and cultural discourses.

As I have already briefly discussed, it is important to note that the author, user, and pirate have not existed as part of a stable holy trinity across copyright's history. These subjects have emerged and receded at different points and have been viewed in different ways depending on the specific social, cultural, and legal contexts in which they operated at the time. In short, we cannot simply assume that the author, user, and pirate are constantly present at every point in the history of copyright law. Therefore, although I will be examining all three subjects throughout this book, at some points one subject will be the focus and another may disappear. At other times, I will narrow my analytic lens to examine points of tension between two subjects, such as the author and user or the user and pirate. This is a feature (rather than a flaw) of taking this relational approach. By attending to the presence or, indeed, the absence of each subject, we can begin to ask interesting questions about copyright law: At what point do particular subjects emerge and others recede? When do the boundaries start to blur between subjects and when do they get shored up? It is only through this sort of critical attentiveness that we can start to seriously unpack and examine subjectivity in copyright law.

Notes on an Interdisciplinary Methodology

The relational framework is interdisciplinary in its approach. Whereas existing relational approaches to copyright law work "within the copyright system,"[81] this framework incorporates cultural analyses of subjectivity into its analysis. As the above discussion of subjectivity has highlighted, cultural studies has a long history of engaging with the "relational as opposed to essential quality of subjectivities."[82] Thomas Streeter notes that it is this very disciplinary history that makes the field a vital space from which to engage with "the dynamics of various constructions of selfhood in specific contexts."[83] Kathy Bowrey and Jane Anderson echo this stance, explaining that the cultural studies project brings a "heightened level of awareness and engagement with issues of subjectivity and the authority of the speaker."[84] In short, cultural studies already attends to subjectivity and embraces its relational qualities as a default disciplinary position.

Moreover, because the study of subject formation also demands an attentiveness to discourse, my interdisciplinary approach also allows for the study of culture to be incorporated into this deployment of the relational triad. This is not a novel development in legal scholarship more generally, but it is worth noting because in some areas of copyright scholarship doctrinal analysis of law still dominates. Therefore the book will also deploy discourse analysis, an approach that has a long history within cultural studies. The discipline considers language to be the building block of culture and the process by which subjectivity is shaped. It thus makes sense to attend to how people and institutions talk about copyright law and about the types of subjects it addresses. Through these discussions, we can also see the relational element of copyright law emerge, with particular narratives showing how actors navigate between the three available subjects or indeed operate within the interstices of this framework. The analysis will also address formal arenas of law. Language is no less prominent in courtrooms, parliaments, or legislation, and, indeed, it is in these spaces where the discursive construction of different subjects is often the easiest to locate.

It is worth noting that despite the interdisciplinary history of intellectual property scholarship, the recent move toward cultural studies is still somewhat uncommon in some academic legal forums. As Cohen explains, "most scholars have assumed that a grand theory of the field must be grounded either in a theory of rights or in a theory of economic analysis."[85] She argues that most legal scholarship still fails to seriously engage with literature from the humanities and social sciences. Madhavi Sunder describes the field in a similar fashion, noting that the majority of legal scholars still tend to frame their analyses of intellectual property law in an instrumentalist fashion by simply critiquing the law's obstruction of further innovation, which they see as the central goal of intellectual property law.[86]

Pushing back against these scholarly tendencies, both Sunder and Cohen call for a greater engagement with cultural studies in the near future. Anupam Chander and Sunder go as far as to demand "a new cultural studies of intellectual property law."[87] *Authors, Users, and Pirates* fits within this wider project, embedding cultural research on copyright law into the wider legal debate around intellectual property. Indeed, it is only though this mixture of disciplines, where the complexities of culture are attended to and the processes, procedures, and histories of law and legal interpretation are taken seriously, that a rigorous analysis of the author, user, and the pirate—and subsequently, copyright law and subjectivity—can take place.

Structure of the Book

In the first chapter we see how the author, pirate, and user emerged from a messy and incoherent process of legal clarification and classification across the eighteenth and nineteenth century. Beginning in the early modern era, I will briefly discuss the historical origins of copyright before discussing an early case around abridgment—*Gyles v. Wilcox*—that stands as an alternate site from which to explore the historical emergence of these subjects. I will then move to the nineteenth century and examine a range of limit cases during this period through which boundaries around infringement were established, contributing to the construction of the author, user, and pirate as legal subjects. I end by critiquing copyright's increased reification of authorship throughout the twentieth century and argue that this process wrongly assumes that subjects operate in separate domains and assigns too much coherence to copyright law.

This historical grounding continues in the second chapter through a consideration of how the user has become an increasingly prominent subject in law over the twentieth century. The chapter will outline the ways in which both law and culture have treated the user, paying particular attention to the emergence of fair dealing in the early twentieth century. I will then examine developments in media distribution and regulation and explore how these systems helped to maintain a hierarchy between users and owners for much of the twentieth century. The chapter then moves into the present day and considers the emergence of the creative user in popular discourse and in legal, media, and cultural studies scholarship. I argue that the status of the creative user underlines the contradictions inherent in copyright law's philosophical foundations and suggest that a relational approach is the best way to think through the tensions produced by this subject.

The book starts to narrow its focus in chapter 3, which is organized around a detailed examination of the relational author. This involves a study of various forms of creative practice, from musical composition to the production of modern art, as well as a consideration of attribution and the system of music licensing. Throughout this chapter, we see how the poorly defined boundaries between authorship, infringement, and use shift and change depending on the type of creative practice in which individuals engage. I suggest that brands might operate as a useful extension of authorship in the context of some creative industries but also discuss various industry practices and organizations that attempt to manage relational interactions between different parties. The presence of these negotiations between practice and law highlights both copyright's occasional struggles with relationality and the limitations of understanding and justifying creative economies solely through the lens of copyright law.

In chapter 4, I undertake a detailed consideration of recent user reforms in the United Kingdom, Canada, and Australia. Through careful analysis, we eventually see that these reform processes reveal the impossibility of maintaining a coherent separation between subjects of copyright law. A critical examination of these reforms underlines the problematic stance of creativity and the uncertain position of the user within law, continuing an argument begun in chapter 2. I finish the chapter with a sustained discussion of recent judicial decisions on new distribution technologies in the US, Australia, and Canada. These cases outline how intermediaries are understood in relation to the user subject, highlighting both the piratical capacities present in the user and the productive potential of consumptive use.

Chapter 5 turns the book's critical lens toward the pirate and outlines the relational aspects of this contested subject. Through a careful consideration of recent scholarly analyses of piracy and an examination of the protests surrounding the Stop Online Piracy Act and the Protect Online Piracy Act, I outline the various slippages that occur around the pirate and suggest that there is significant overlap between the pirate and the user. The chapter then takes a wider transnational view by exploring how piracy functions in postcolonial countries, revealing the authorial capacities present in piracy.

This chapter contends that although the pirate is often institutionally and discursively constrained, any act of piracy will reveal both an element of "use" and a latent authorial capacity.

These theoretical claims about the pirate are further developed in chapter 6 through a detailed examination of how the pirate has been considered in landmark cases so far in the twenty-first century. I begin with a brief contextual discussion of the first major digital piracy case: *A&M Records, Inc. v. Napster, Inc.* I then return to Australia where some of the more prominent cases around piracy and online regulation have occurred: *Universal Music Australia Pty Ltd. v. Sharman Networks Ltd.* and *Roadshow Films v. iiNet Ltd.* I draw on the evidence of the court to show how cultural meanings and media practice play out in legal forums. After assessing these various discourses around the pirate, I examine the cultural work law needs to undertake in order to maintain a clear separation between these various subjects.

The book concludes with a summary of the proposed relational approach and discusses its implications for law, policy, and scholarship more generally. I suggest that a relational approach to copyright law and subjectivity offers a productive conceptual method to use as the implications of various legislative, technological, and social changes around copyright law are debated. As part of this concluding point I also reiterate the importance and value of undertaking interdisciplinary research on copyright law.

1 Connected Origins: Locating Relationality in Copyright History

This chapter contends that the subjects of copyright law have always been relational. In the following sections, I outline the emergence and subsequent development of copyright law across the eighteenth and nineteenth centuries in the UK and show how the absence of coherent jurisprudence has allowed relational subjects to emerge.[1] After beginning with a brief background on the history of copyright, I discuss the issue of abridgment to provide an example of the co-emergence of the author, user, and pirate. Competing decisions around how much of an existing work could be taken before being considered infringement contributed to the continual reorganization of relationships between the author, user, and pirate during this period. The chapter ends with a short reflection on the increased focus on authorial rights that typified twentieth-century copyright policy and (perhaps surprisingly) argues that despite the instrumental approach to authorial rights taken in this period, these three subjects still interacted relationally with one another.

At the outset, it is worth noting that the chapter also offers a very brief history for those who are unfamiliar with copyright law and functions as a historical background for the chapters that follow. Subsequently, the chapter occasionally skims over complicated cases, covers three centuries relatively quickly, and does not offer novel historical findings. Still, it offers a novel perspective on existing cases and scholarly narratives by identifying the historical presence of relationality in copyright law, tracking the emergence of the relational triad, and identifying how it changes over time. It also contributes to a wealth of work that has identified a lack of substantive conceptual development in copyright law, showing how this absence of theorization has allowed for a relational notion of subjectivity to emerge over the centuries.[2]

Enacting and Interpreting the Statute of Anne: Authors, Intermediaries, and the Market

If we are to examine subjectivity and copyright law seriously, we first need to explore the origins of copyright and examine how the author, user, and pirate were constituted during the early days of this legal framework. This process shows how both the cultural constitution of these subjects and the market contributed to the construction of legal subjectivity and allows us to sketch out the initial relational position of these three subjects. The first copyright law was the Statute of Anne, which was enacted in the United Kingdom in 1710 and was the first time that authors were ostensibly recognized as owners of their works.[3] The statute provided a term of protection of fourteen years, renewable for another fourteen years. As we will see, however, this law was an attempt to manage an ongoing regulatory dispute within the printing industry and so did not provide much legal power to authors.

Prior to the statute, the Stationers' Company, a trade guild for all workers associated with the production of manuscripts, had been granted an effective monopoly over printing in England (and later Britain). This was due to the Stationers' Charter, which gave the company "powers of national regulation" over the printing trade in 1557 and embedded this publication system within a functioning censorship regime.[4] But a growing anti-monopoly sentiment among the English public as well as the political turmoil caused by the English Civil War gradually weakened this arrangement. The regulatory system collapsed in the mid-seventeenth century during the war and was not restored until the monarchy returned to power. Although the Crown returned the general monopoly on printing to the Stationers' Company through the Licensing Act of 1662, the act was "to remain in force for only two years," which meant that it needed to be renewed regularly.[5] The aforementioned anti-monopoly sentiment eventually grew to such an extent that Parliament did not renew the Licensing Act in 1695, once again eroding the monopoly power of the Stationers' Company.

The House of Commons argued that the act was inefficient and subject to favoritism and abuse.[6] The learned public had also started to resent the control that a small cartel of publishers retained over a number of important texts as well as the imposition of prepublication licensing and censorship.[7] The free press of England, which dramatically expanded at the turn of the eighteenth century, was also beginning to be valued seriously by younger members of Parliament, who were developing a "taste for uncensored news and information."[8] With prominent figures like John Locke and Daniel Defoe arguing for freer presses and "an increased emphasis on post-publication accountability and prosecution," it was unlikely that a bill that simply echoed the 1662 Licensing Act would be passed in the near future.[9]

What is interesting for our purposes is the extent to which the Stationers' Company gradually made use of the authorial subject in its attempt to retain guild control over printing. Because the Company functioned as part of the wider censorship regime in Great Britain, it originally attempted to restore its old monopoly by linking itself with pro-censorship forces.[10] Nonetheless, the decade following the 1695 lapse of the Licensing Act saw thirteen proposed bills on press regulation fail and the Stationers' Company eventually heeded the winds of change.[11] It gave up the dream of maintaining a monopoly on the printing industry and acknowledged it might have to concede some ground to anti-monopolists in exchange for the viability of the industry as a whole.

A discursive change in the Stationers' claims occurred in 1706. They no longer highlighted "their precarious livelihood and the numerous stationers' widows facing utter ruin."[12] Instead of focusing the need for censorship on worries about their own financial wellbeing, the Company framed its cause around its strong support of authors' rights and the need to sustain regulatory systems that allowed knowledge to be disseminated. This change in rhetoric was an important moment in the emergence of the author as a legal subject and the notion of subjectivity in copyright more generally. The Stationers were suggesting for the first time that authors, in addition to printers and vendors, had a legitimate "claim upon the legislature's attention."[13] This shift was successful to the extent that by 1709, "support for some form of legislation regulating the printing industry was widespread throughout the book trade."[14] The result of this lobbying was the first copyright law, the Statute of Anne.

It is important to note that even though authors were granted rights in law, the passing of this legislation did not immediately change practices of ownership or distribution in early modern England.[15] The statute represented a significant shift away from monarchical regulation and hinted at our modern configuration of copyright law, but it was not entirely clear how this new statute would actually work. Indeed, the only reason the Stationers were happy that Parliament enacted the Statute of Anne was because they assumed that a statutory right would have little to no impact on the operation of trade. The Stationers acknowledged that authors had a statutory right to their work but believed that it would be transferred to them through contract. Furthermore, although they held that authors retained a perpetual common law right of property to their work, because authors needed to sell their works to printers in order to get published, this common law right was similarly transferred to the Stationers.[16] At the point of enactment, the Stationers only had to deal with two major changes: first, the fact that the statute was focused solely on trade regulation and "severed" the censorship process from the printing trade and, second, the opening up of the printing trade to "non-members of the Stationers' Company."[17]

Subsequently, at copyright's birth subjectivity was already bound up in a number of complex arrangements that would continue to affect this legal framework in the years ahead. The most obvious issue to come out of the preceding narrative is that the author was essentially functioning as an avatar for a regulatory claim made by the central actor in the printing industry, the Stationers' Company. This is not to dismiss the valid claims for authorial rights that had been occurring across the centuries but rather to note that the historical record shows this initial claim to authorial rights was pushed through strategically by intermediaries rather than by the authors themselves. Subsequently, the Stationers' Company positioned the author as a cultural and legal subject, with the author used as a convenient avatar through which existing trade practices could be maintained. The Stationers conducted this discursive work under the presumption that authors would align themselves with this new legal regime, with little promise of gaining much in the way of new rights, and this episode demonstrates the initial importance of intermediaries and market considerations in the constitution and construction of the author in copyright.

It is important to qualify this narrative of authorial emergence in the context of the relational triad proposed by this book. The Stationers merely assumed that their strategy of using the author as an avatar for their attempt to reinscribe existing trade practices would work. In reality, it was not yet clear how the statute would operate in practice. Copyright law was fundamentally undetermined at this point. There was no clear direction around the remit of the authorial right, exactly what it protected, whether elements of the protected text would be available for use, or indeed the extent to which penalties would be enforced. Therefore, at this early stage when addressing the legal subject, we can view the author, user, and pirate as essentially bound up together as proto-subjects of copyright. Contrary to scholarly narratives of authorial emergence suggesting that the construct of the author was "introduced to English law in 1710" as a "charged receptacle,"[18] the enacting of the statute resulted in a series of connected, legally dormant subjects who would emerge collectively over time.

Alternative Historical Sites of Subjectivity: Relational Subjects and Judging Abridgment

The issue of abridgment offers a useful example of the relational emergence occurring soon after the Statute of Anne was enacted. Although the Stationers' Company occasionally attempted to control the production of derivative works prior to the statute,[19] there is evidence of a series of abridgments being freely published with little comment from the Stationers in the seventeenth century.[20] The "emergence of new markets for different and cheaper literary materials" in the early eighteenth century, however, pushed the booksellers to look for a way to control markets for derivative works.[21] They first attempted to push through a bill that addressed the practice of abridgment and when that failed, they brought a series of cases before the Court of Chancery.[22] A series of cases around this issue were heard in 1739 but the issue of abridgment was not clarified until 1740 in *Gyles v. Wilcox*.[23]

This case required Lord Hardwick to adjudicate on whether the law book *Modern Crown Law*, to be published by John Wilcox, borrowed "verbatim" from *Matthew Hale's Pleas of the Crown*.[24] Hardwick noted that "colourably shortened" books would fall within the meaning of the statute. He believed, however, that the statute should "receive a liberal construction, for it is very far from being a monopoly." In turn he offered a principle of abridgment with reference to the statute, claiming that "abridgements may with great propriety be called a new book, because ... the invention, learning, and judgment of the author is shewn in them."[25] Hardwick then directed two people "skilled in law" to conduct a close reading of the two books, and they found that *Modern Crown Law* was a fair abridgment.[26]

The *Gyles v. Wilcox* decision, occurring only thirty years after the statute's enactment, offers an isolated example of how courts were approaching the three subjects of copyright law relationally. Although Hardwick's interpretation of abridgment appears to stand as a precursor to discussions around the scope of particular forms of "use" authorized under copyright law, it can also be read as significantly extending the authorial remit provided by the statute. A literal reading of the Statute of Anne limits only the wholesale copying and distribution of manuscripts. Subsequently, it could have been construed that any partial copying of works did not fall foul of the law, which would have meant that a prospective user of copyrighted material could operate with only a limited remit for the author (and consequently, the pirate), an interpretation of the statute that Jane Ginsburg describes as "a little coral reef of private right jutting up from an ocean of public domain."[27] In contrast, Hardwick's interpretation expanded the authorial remit offered by the statute by finding that abridgments that did not show substantive evidence of authorial contributions by the second party would also be viewed as infringement.

A series of relational conceptualizations of legal subjectivity can also be identified in Hardwick's decision. A relational approach to authorship appears, with use being presented as an act that a prospective author can engage in to produce a new work. This gestures toward the relational and interstitial subject of an author–user by recognizing that future acts of authorship must draw on previously published works. We also see an alternate vision of infringement appear with an infringer positioned as an individual who does *too little* with the primary work. Although the user has not yet emerged as a dominant cultural subject, prospective pirates and authors are directly linked to the act of use and are constituted in relation to this act.

Hardwick's interpretation of abridgment also stood in opposition to cultural depictions of piracy that were prominent during the early eighteenth century. Ronan Deazley notes that during this period abridgments were regularly referred to as piracies by authors, who viewed them as simply unauthorized reprintings.[28] Daniel Defoe, in his *Essay on the Regulation of the Press*, argued that when "An Author prints a Book ... it shall be immediately abridg'd by some mercenary Bookseller, employing a Hackney-writer."[29] This view positioned abridgment as a wholly piratical and non-authorial act and contributed to broader cultural imaginaries of authorship and piracy. Although these acts (and the subjects associated with them) were still being developed as legal concepts, these narratives show that the author and pirate were already in operation as cultural subjects. The stark contrast between authors' descriptions of abridgment and Hardwick's own judgment, however, reveals that in particular legal cases, copyright was developing a vision of these acts operating in a relational fashion.

At this point it is worth noting that *Gyles v. Wilcox* cannot tell us much about the construction of subjectivity in copyright law more generally. This is because there was not yet a coherent culture of jurisprudence in what was a fundamentally disorganized legal system. The United Kingdom's legal culture during this period was radically different from modern jurisdictions with "numerous approaches to legal reasoning and techniques in operation ... and competing views on the relative authority of legislation over common law, and common law over equity."[30] There was also an absence of legal theorization. For example, "courts could reach results in property cases without any clear theory in mind."[31] Therefore, although a cultural vision of the author and the pirate may have been somewhat apparent and we can see faint sketches of the author, user, and pirate, this case can only stand as a particular instance of the development of these subjects due to the lack of jurisprudence guiding the decision. The decision is also somewhat limited in its historical impact, with *d'Almaine v. Boosey* in 1835 constraining the scope of abridgment by "denigrating the quality of the adapter's authorship."[32]

I suggest, however, that *Gyles v. Wilcox*[33] offers an alternative way to think through the development of subjectivity in the early period of copyright's history. A significant amount of scholarly attention has been paid to the "Battle of the Booksellers," where two cases (*Millar v. Taylor* and *Donaldson v. Beckett*)[34] addressed the central issue that the Statute of Anne raised: did authors, and through them booksellers, have a perpetual common law copyright in their works or were their rights confined to the statutory period provided under the statute?[35] An ongoing scholarly debate has explored the ramifications of *Donaldson* and some have argued that these two cases stood as a significant intervention in copyright's conceptualization of the author.[36] However, *Gyles v. Wilcox*, which is concerned with the possibilities afforded to a proposed activity under copyright law rather than the duration of the right, also offers a productive site from which to examine how the courts thought about emerging (but as yet undefined) legal subjects.

Moreover, *Gyles v. Wilcox* offers an early example of the problems and limitations around the constitution of legal subjectivity in copyright, issues that are still present today. Hardwick interpreted the statute consequentially and suggested that he could not offer a strict interpretation of abridgment as this "would be of mischievous consequence, for the books of the learned" would immediately be viewed as infringement.[37] Therefore, Hardwick found that some abridgments must be allowed under the statute. As noted previously, the decision affected the functional constitution of all three subjects in a relational fashion. In an early example of an ongoing tendency in copyright jurisprudence, however, Hardwick did not engage in a substantial theorization of any of these subjects.[38] Therefore, although his consequential reasoning in judgment signaled that subjectivity is produced relationally, there was little attempt to establish the broader conceptual underpinnings of each subject.

The next section continues this historical exploration of the relational triad and subjectivity by examining a series of nineteenth-century court cases around abridgment. The nineteenth century presents a series of cases with no central jurisprudential line running through them. Their complexities and contradictions provide excellent examples of relationality, with the agencies of each subject overlapping regularly over the years across these divergent judgments.

Copyright Law in the Nineteenth Century: Debating Use and Infringement in the United Kingdom

The relational nature of subjectivity is easily located in a range of cases throughout the nineteenth century because copyright was still undetermined. Judges continued to offer different interpretations of infringement and use depending on the case, with subjects constantly defined and redefined through comparison with each other. This occurred directly in infringement claims, with the "claims to authorship of both plaintiff and defendant" assessed "side by side" rather than through consideration of the "extent and justification of the exclusive right of original author."[39] The lack of legal clarity also meant that judges continued to separate infringement from authorship and use in idiosyncratic ways, which produced varying constellations of the author, user, and pirate. These uncertainties tell us about the cultural status of these subjects as well as their status as legal subjects. During this period it was possible to locate an author or pirate but, in marginal legal cases where claims of authorship were at their weakest, the lines between the author, user, and pirate still blurred. Was the compiler of a postal directory, who used information from an existing work, an author and user (of freely available facts) or a pirate? At this point in time, neither law nor culture was entirely sure.

The 1826 case *Mawman v. Tegg* offers a good initial example of judicial reflection around these connections.[40] Mr. Mawman and twenty associates accused Mr. Tegg, a notorious cheap reprinter, of replicating sections of the *Encyclopaedia Metropolitana* (in which they held an interest) in his work the *London Encyclopaedia*, arguing that "many of the articles ... were copied verbatim, or nearly so."[41] Tegg's defense rested on five points of contention. First, it noted that the two works were structured differently: the *London Encyclopaedia* was alphabetical whereas the *Metropolitana* was "distributed on philosophical principles into four divisions."[42] Second, it suggested that there would be little market impact on the *Metropolitana*, as the works were "intended for two classes of readers totally different." Instead of being published in "large and expensive quartos," the *London* would be released in octavo volumes at one third of the *Metropolitana*'s price.[43] Third, the affidavit argued that much of the existing material was "taken from the lexicographical department" and that "surely any man was at liberty to compose a dictionary of the English language."[44] Fourth, the replicated text "formed a very inconsiderable part of the whole publication." Finally, the defense alleged that the *Metropolitana* had also borrowed "from other works."[45]

Lord Eldon, who adjudicated the case, noted that it was quite unique. Previously, the cases he had decided on had

all been either cases, in which, by altering or destroying the title-page, the publication could go on, or where the parts, that were pirated, bore such vast proportion to the whole of the work, that there was no difficulty in ascertaining, whether the work was or not upon the whole piracy, or piracy to such an extent that you could feel no uneasiness in granting an injunction.[46]

Lord Eldon also reflected on a judgment of Lord Ellenborough's in *Cary v. Kearsley*, which stated what had since become an oft-cited formulation of how existing copyrighted works should be used.[47] Echoing Ellenborough, Eldon felt that it was important that the matter being copied had been used to "give to the public what might fairly be called new work," but that one could not take too much and in turn act *animo furandi* (with intent to steal).[48]

Though the issue was settled out of court with Tegg paying the plaintiffs "a considerable sum of money," the case is notable for Eldon's judicial articulation of the boundaries between the pirate, the user, and the author.[49] It was possible to use copyrighted material if one was found to have authored a new work; otherwise, that individual would risk becoming a pirate.

In contrast, the 1866 case *Kelly v. Morris* saw the boundaries between authorship, use, and infringement firmly separated, with the decision moving away from the formulations of Lord Ellenborough and Lord Eldon.[50] Beginning in 1836, Frederick Kelly published the "Post Office London Directory" every year, with the directory split into three sections: a street directory, a commercial directory, and a court directory. In 1866, Mr. J. S. C. Morris published "The Imperial Directory of London," which resembled the 1865 edition of Kelly's directory to the extent that "errors were discovered which were common to the two works."[51] In his defense, Morris, who had previously published his own Business Directory, argued that his new Imperial Directory had corrected some errors found in Kelly's directory and featured additional names, "particularly in the new parts of London."[52] His method of compilation was to make lists of names and addresses from various sources, which he admitted included the 1865 edition of the London Directory. However, Morris then said that he had sent out canvassers to correct these lists where they were inaccurate, and also insisted that "persons who had allowed their names and addresses to be printed in a directory had made them public property."[53]

In deciding the case, Vice Chancellor Wood found that the Imperial Directory was not a new work, and he placed an interesting emphasis on labor in his decision. He stated that Morris had not found the information for himself but had instead "copied the plaintiff's book first" and then gone to see if the details were right.[54] In closing, Vice Chancellor Wood declared that "it has been clearly made out that the defendant has not compiled his book by legitimate means and from his own personal labour, or that of his agents, but has made great use of the plaintiff's book."[55] He also presented a stricter interpretation of infringement than many of his contemporaries, remarking that the defendant was not entitled to "a single line" or "one word" of the London Directory if the requisite labor was not performed, thus underlining the uneven conceptual status of infringement during this period.[56] Wood's judgment presented entirely new trajectories for authorship, use, and infringement, with authorship directly tied to labor, use almost entirely absent, and the notion of infringement expanded to include the reuse of what many would have considered factual information.

Wood's decision was echoed in 1868 in *Morris v. Ashbee*, a case that concerned Morris's earlier publication: the Business Directory of London.[57] This time, Morris was the plaintiff instead of the defendant and was accusing former employees Mr. Ashbee and Mr. Simmonson of producing "a piracy of the Plaintiff's work in general form and arrangement" that was also "to a great extent printed from slips cut from the last year's edition of the plaintiff's directory."[58] Vice Chancellor Giffard found for the plaintiff and similarly emphasized the role that labor played in his decision, noting that unlike Morris's Imperial Directory, the Business Directory of London clearly involved substantial effort on Morris's part. Vice Chancellor Giffard explained that "the plaintiff incurred the labour and expense first of getting all the necessary information for the arrangement and compilation of the names as they stood in his directory."[59]

These debates around the extent to which an individual could claim authorship over information and the role of labor in authorship continued in two judicial decisions during the 1890s. In *Collis v. Cater*, the court found that there was "copyright in a list of articles" and that a man had no "right to appropriate to himself without payment or recognition in any way what it has cost his neighbour expense and trouble to make out."[60] A distinction was made between public information and works emerging out of the labor of an individual, with the court noting that an individual "incurs [a] good deal of trouble ... in preparing [a] full catalogue such as either of these works have."[61] Lord Herschell reached a similar decision in *Leslie v. Young* and, due to the particulars of the case, was able to position a clear judicial line between public information and works deserving of copyright protection. He decided that the collection of "time-tables which are to be found in railway guides and the publications of the different railway companies could not be claimed as a subject-matter of copyright."[62] Still, he stated that a compilation of useful information about "circular tours" in an area displayed sufficient labor to be copyrightable.[63]

Across these cases a trend emerges with the judiciary gradually valuing authorial labor along with "competition and the plaintiffs' claims to an expanding market" when assessing infringement claims.[64] These trends should not, however, be read as part of an identifiable jurisprudence operating throughout the nineteenth century.[65] Instead, during this period we see judges offering competing and occasionally contrasting positions on how to define use, authorship, and infringement. Authors, users, and pirates are placed in a range of different relationships and the boundaries between each figure are often redefined and rearticulated as courts are faced with different facts. In one sense, these processes simply reflect the comparative freedom of legal interpretation that was afforded to the judiciary during this period thanks to the "disorderly and unmethodical appearance of the common law system."[66] But they also highlight the extent to which these three subjects were co-constructed and defined in relation to one another. The shifting boundaries between each subject that occurred in different judicial interpretations signals the nonobvious nature of these subjects and the absence of any one clear legal or cultural concept animating these subjects across the century, particularly around the limit cases discussed above.

It is also worthwhile to note that despite the absence of a consistent jurisprudential logic driving the process of subjectification, the ideological construction of subjects was still present throughout. We can look at the attention paid to authorial labor, which gradually comes to define authorship, the lack of labor that seems to point to evidence of infringement in particular cases, or, alternatively, the opportunities made available for use (e.g., when considering the factual and ostensibly public nature of railway timetables). In each of these examples, judges were asking, "What does authorship, use, and infringement look like?" and, in doing so, were offering an ideological picture of each subject and their role within copyright law. Indeed, it is through this process that the concept of a pirate under copyright law transitioned across the eighteenth and nineteenth century from a wholesale copier of a physical book to an individual who may have inappropriately copied a selection of words. Still, due to the relational nature of subjectification as well as the uncertain legal context in which these decisions took place, this picture of the relational triad is always undetermined and subject to change. In the final section that follows, we see what happens to these relational interactions as copyright law is finally defined (and interpreted) through legislation.

Authorial Dominance and the Complications of Copyright

An entirely new copyright act was enacted in the United Kingdom in 1911 that addressed some of these jurisprudential inconsistencies. I will not discuss the act in detail but I mention it because it was the first time that infringement and use were defined and codified in UK law. As a result, the "boundaries of infringement" were altered as the authorial right expanded to include "translations, performances, dramatisation and mechanical reproduction."[67] The right to abridgment also became part of the author's remit, with the question of whether the "new work" brought any benefit to the public no longer seen as relevant. The use of copyrighted works was placed under a similar set of constraints with many of the commonly allowed uses (such as quotation for the purposes of review) now specifically detailed in a set of "fair dealing" principles (discussed further in chapter 2). At the time of the legislation's enactment, it was presumed that these exceptions merely codified existing law, but these principles were gradually interpreted in a relatively strict fashion.[68] Indeed, as Kathy Bowrey notes, although these attempts at codification were seen to be relatively uncontroversial, the cumulative effect of this legislation was to formalize an author-centric statutory law, with "defendant's work[s] now only appear[ing] in the context of defences to infringement."[69]

An increased doctrinal focus on the author continued throughout the twentieth century through a combination of cases and legislative amendments in the United Kingdom as well as other common law countries (including the United States), which resulted in an ongoing expansion of the authorial right.[70] For example, the definition of what constitutes a copyrighted work expanded over this period. In 1802, the first amendment to the 1790 US Copyright Act extended the right to prints and etchings[71] and the 1831 amendment granted copyright to printed music.[72] Lyman Ray Patterson and Stanley Lindberg concisely outline the further extension of copyright in the United States:

[D]ramas in 1856, photographs in 1865, the entire panoply of artworks in 1870, the performance right for musical compositions in 1897, motion pictures in 1912 and sound recordings in 1972.[73]

To this list we can add another notable US amendment that incorporated software into the Copyright Act as a copyrightable work in 1980,[74] as well as the extension of copyright to useful articles that "have both a utilitarian function and aesthetic design features such as belt buckles, ashtrays, and coffeepots."[75]

The term of copyright protection has also been extended over time. To return to the UK for an example, whereas the Statute of Anne granted authors only fourteen years of copyright protection with a possibility of another fourteen years following renewal, the Copyright Act of 1842 granted protection for the extent of the author's life plus seven years following their death, or forty-two years of protection (whichever was longer).[76] The Copyright Act of 1911 extended this to fifty years after the death of the author and the term currently stands at seventy years after the death of the author.[77] Similar extensions have been granted in comparable jurisdictions, with the United States' copyright term also extending from the original (once renewable) fourteen-year term to seventy years after the death of the author.[78] This increased protection has also become internationalized and entered the realms of trade policy, with a range of international agreements gradually ramping up copyright protection and enforcing this authorial dominance in copyright policy on a global scale.[79]

The turn toward authorial protection across the previous century was relatively pronounced. This was largely due to a fervent belief in the inherent logic of copyright, a belief commonly held by certain policymakers and lawyers in some creative industries who argued that copyright was a simple law. The author was easy to locate (as are acts of authorship) and deserved protection, users were given relatively short shrift in terms of their rights under the law, and infringement was easy to find and an ever-present threat. These advocates for authorial expansion positioned subjects as distinct and disconnected actors—authors were authors, users consumed (they did not produce), and infringers were infringers. Their view was a significant shift from the diverse assessments of these three subjects given throughout the nineteenth century, which as we saw were occasionally open to relatively nuanced readings of the balance between the author, user, and pirate.

Scholars have critiqued the views of these advocates, often called "copyright maximalists," largely because they failed to recognize the publicly oriented rights and outcomes of copyright law. I, however, want to focus on how these maximalists had to grant copyright a level of internal coherence that it does not have in order to establish an ideology that separated their author (and ideas of authorship) from other subjects of law. For example, a simple description of a basic copyright principle is that expressions can be copyrighted whereas ideas cannot. Maximalists could make the case for increased authorial protection because the realm of ideas is always open to the prospective user to draw from. Still, this separation between the unique expression of the work and the general idea that animates it has never been entirely fixed. Christopher Buccafusco asks the obvious question, "[H]ow can we even say what a painting by Piet Mondrian or a composition by Arnold Schoenberg is 'about' so we can begin the process of differentiating ideas from expression?"[80] He goes on to note that the uncertainty of this principle has been an ongoing problem and point of discussion among copyright scholars and judges for some time.[81] Similar issues have emerged around the claim that copyright does not protect systems or processes. Buccafusco points out that software code is copyrightable whereas a recipe is not, despite the fact that both are essentially processes.[82] Therefore, although the twentieth-century authorial expansion recognized more and more authors as legal subjects, with an increasing bundle of rights, it was based on a misrecognition of how subjectivity is constituted within copyright and a presumption of legal coherence.[83]

Indeed, recent copyright scholarship has outlined how this twentieth-century model of authorship could only function by moving from the site of the subject and toward the realm of the marketplace, with "the author" often standing in as a heuristic device in order to support the introduction of emerging industries (from movies to computer software) into the copyright pantheon.[84] Subsequently, the author as subject was emptied out as a substantive category in order to effect these changes in copyright law, with the recognition of new subject matter often based on the judiciary or the legislature identifying minute differences between acts of protected and unprotected authorship rather than engaging in substantial conceptual work, which resulted in the kind of contradictions outlined by Buccafusco. Much like the Stationers' Company at the start of this chapter, a range of industry intermediaries mobilized the author as a market device and "alienat[ed] [many] creative workers from the products of their labor."[85]

This dominance of the author in copyright doctrine throughout the twentieth and twenty-first centuries has been read productively as evidence of the diminished role of the user as a legal subject.[86] This position is eminently arguable. The expansion of the authorial right across this period has limited the scope for use, meaning there is subsequently more risk of being identified as an infringer. Still, it is important to keep in mind that although these transformations can impact the legal and cultural constitution of subjects, these maximalist approaches are based on a limited understanding of authorship.[87] Therefore, even though the effects of "maximalist" copyright policy should rightly be examined and discussed, we cannot read too much into it as a model from which to understand subjectivity because it presumes that copyright law emanates from a solid conceptual base.

The chapters that follow explore how this twentieth-century approach to copyright has affected the cultural and legal constitution of all three subjects well into the twenty-first century. We also see stronger evidence of a progressive form of relationality occurring, with various socio-legal discourses directly influencing a relational constitution of legal subjectivity, protest groups redefining their own subjectivity under copyright law, artistic practices complicating notions of authorship, and courts approaching the pirate with a nuanced relational approach. The incoherence of copyright ultimately produces situations where use and infringement are not always meaningfully distinguishable from authorship, particularly as authorship increasingly functions as a market device rather than a coherent category.

Despite the introduction of various legislative instruments that aim to clarify case law and a dominant maximalist narrative that argues for the coherence of copyright law, copyright is still somewhat underarticulated and contradictory. Subsequently, with authorship or the idea/expression dichotomy unable to be logically discerned on occasion (to give just two examples), it follows that all three subjects will continue to be relationally connected, since legislation and jurisprudence cannot always meaningfully separate them. This linkage occurs at a basic level (in the sense that all three are related legal subjects) and allows these three subjects to overlap, change, and interconnect in response to legal and cultural developments over time. The relational nature of copyright is evident in the early years of copyright's history and the fact that it has continued through the twentieth and twenty-first centuries (which the following chapters will explore in detail) refutes the broader ideological claim made by copyright maximalists that copyright is a simple site from which to discern and separate subjects.

This chapter has made a case for viewing the three central subjects of copyright law as meaningfully connected from the birth of copyright onward. It deployed this relational triad in an analysis of *Gyles v. Wilcox*, detailed the lack of coherence around these subjects in the nineteenth century, and discussed the extension of authorial rights that occurred across the twentieth century. As noted at the outset, I have had to elide or truncate some notable cases, long narratives of legal reform, and other important social and cultural trends in order to outline this broad historical process. In doing so, however, we have been able to clearly see how competing principles and philosophies of authorship, use, and infringement regularly changed the legal and social remit of the author, user, and pirate. In the next chapter, we turn to the user directly and deploy a relational analysis to examine its development as a legal and cultural subject.

2   From the Piratical to the Creative User

The subject of the user has undergone a significant transformation. This chapter tells the story of this transformation and examines how the user's relationships with other subjects of copyright law changed across a long century (1911–2017). From 1911 onward, the user became gradually linked with the pirate thanks to the expansion of the authorial remit discussed in chapter 1, the presence of a system of media regulation that separated users and owners, and narrow judicial interpretations of copyright exceptions. As various disruptive technologies were progressively domesticated, however, cultural and legal conceptualizations of the user shifted until the subject started to be seen as creative in its own right. This study of the user is an example of how regulatory frameworks, cultures of use that emerge around media, and the progress of cultural industries (among other things) simultaneously contributed to the development of the user and the other subjects under copyright law's remit.

Moreover, the chapter stands as an example of how a relational approach to copyright law can be used to closely examine specific subjects and consider how various ideologies affect their subjectification. The following analysis outlines the agencies associated with the user, examines how law, technology, and cultural discourse interact with and interpret these agencies, and goes on to situate these developments in relation to the author and the pirate. This study reveals that the user is often viewed as an intermediary subject that develops coherence as a legal subject by being associated with either the author or the pirate. The relational triad provides a framework from which to develop a more accurate account of the user as a subject whose legal and cultural constitution is directly influenced by concomitant interpretations of the author and pirate.

The chapter begins with a brief explanation of how the user has been constituted through fair dealing exceptions, fair use exceptions found in the US, and various limitations to the authorial right. I then show how the user's limited position in law can be fully understood only as part of a wider regulatory and technological context through a detailed examination of the development and regulation of radio, television, and video in the UK and US. Advances in media technology during the mid- and late twentieth century begin to give this subject more cultural agency. We then move to the twenty-first century and explore how new discussions of participatory culture have injected an authorial strain into the user, and also examine how copyright law and, more specifically, fair use jurisprudence both contributes and responds to these changes.

The User in the Commonwealth: A Story of Exceptions and Limitations

The user has largely been ignored in jurisprudence.[1] Claims that the user has been entirely disenfranchised are, however, somewhat overstated. As we saw in the previous chapter, questions of use have been present throughout copyright law's history and have directly shaped the development of copyright. There are also exceptions, which allow the use of copyrighted material in modern copyright law, as well as broader provisions and limitations to the authorial right that similarly address the issue of use. The question then is not whether the user and acts of use have been addressed in law (at a basic level, they have) but rather how does copyright law actually understand the user?

There are a number of exceptions to copyright infringement found in Canada, the United Kingdom, and Australia that rest under the title of "fair dealing." Considering that the concept of fair dealing originated in the UK, I will primarily focus on how these exceptions developed there. As briefly noted in chapter 1, fair dealing was introduced in the Imperial Copyright Act 1911 and a range of situations were identified in which a particular act would "not constitute an infringement of copyright."[2] An individual was able to draw on copyrighted works "for the purposes of private study, research, criticism, review, or newspaper summary."[3] If a person authored an artistic work but did not own the copyright, they could still use "any mould, cast, sketch, plan, model, or study [they] made for the purpose of the work," as long as the main design of the work was not repeated.[4] Sculptures, artistic works, and architecture were also not protected from being painted, photographed, or otherwise replicated if they were "permanently situated in a public place."[5] Other provisions allowed sections of copyrighted literary works to be used for educational publications, and the "reading or recitation in public by one person of any reasonable extract from any published work."[6]

These broad guidelines attempted to codify the balance between use and protection of copyrighted works that had developed haphazardly over the previous two centuries. In codifying these actions, however, the new fair dealing provisions (most probably unwittingly) set out a much more restricted scope of action for the user than much of the divergent boundary setting of the past. The Copyright Act 1956 (UK) entrenched these limitations further by directly changing the language of the law. This was noted in the mid-1950s by the Australian Copyright Law Review Committee, who were deciding which amendments should be "incorporated into the Australian copyright law":[7]

The nature of copyright as created by the 1911 Act was defined by specifying the sole rights, which by virtue of the Act accrued to the owner of copyright. In the 1956 Act copyright is defined as the exclusive right to do and to authorize other persons to do certain acts, which are designated as "the acts restricted by the copyright." This seems to be a somewhat cumbersome and misleading method of describing the rights which copyright bestows on its owner. It directs the mind to the infringer—to the things which must not be done without the owners' consent—rather than to the owner and what is comprised in his ownership.[8]

This reorientation shifted the position of the user drastically. Whereas during the uncertain nineteenth century the copyright holder and the user were on something of an even footing, the progressive entrenching of fair dealing in these countries saw the user gradually being positioned as a potential pirate and submissive to the author.

The last major reform in the United Kingdom was the Copyright, Designs and Patents Act 1988. This act made exceptions for: research and private study; criticism, review, and news reporting; and the incidental inclusion of copyrighted material, education, the function of libraries and archives, and public administration, among other areas.[9] Further amendments granted exceptions for the assistance of visually impaired individuals and time-shifting broadcast programs.[10] Similar (but not identical) fair dealing exceptions can also be found in Canada and Australia.[11]

If we reflect on these fair dealing exceptions, it becomes clear that a particular vision of the user is presented. Until a spate of reforms across various jurisdictions in the last few years and recent developments in Canadian jurisprudence (both of which will be addressed in more detail in chapter 4), these exceptions largely assumed that the use of copyrighted content must carry a clear public benefit. The reporting of news, the pursuit of education, reviewing or critiquing existing works, assisting the visually impaired, and contributing to the function of libraries or governance fit in to this category and underline a general assumption of a publicly minded user. It is only in the practice of time shifting where a direct benefit to the wider public (as opposed to the immediate public of a household) might be hard to discern. But what we can say is that in these cases, the available defenses are generally institutional or utilitarian and linked to specific spheres and activities.

In these countries a range of additional limitations and statutory provisions also relate to the user. A prospective user is able to use: ideas, which are unable to be copyrighted; a less than substantial part of a copyrighted text; work that has fallen out of copyright; or material for purposes covered by statutory licenses, pursuant to a fee paid to the copyright holder. Furthermore, once someone has purchased a copyrighted work, they are able to lend it to others through the principle of copyright exhaustion.[12] The provisions for libraries and universities that allow these institutions to reproduce works for archival or educational purposes are also worthy of further mention. These provisions suggest that copyright law views public institutions such as libraries as a representative space for users. With their general public access and ability to offer information for free, libraries stand as spaces where individuals can access copyrighted content easily and at a limited cost.

Exceptions to copyright offer a defined area in which the user can operate. The benefit of this approach is that individuals are presented with a clear set of provisions that are supposedly relatively easy to understand. The downside is that unless legislators have anticipated a use, acts are usually considered infringement. At a practical level, this means that law is unable to quickly respond to technological advances and changing user practices. It is also worth noting that although the exceptions and defenses beyond fair dealing frequently make sense, they are often unreliable defenses that require a detailed knowledge of copyright jurisprudence (such as working out what is "less than a substantial part" of a copyrighted work) or carry high transaction costs (such as licensing music).[13] Therefore, unless the individual is a well-read copyright lawyer, the prospective user has to rely on guesswork and their own (often inexpert) judgment when using copyrighted material.

Fair dealing exceptions position the user next to the pirate as a subject that is always on the defensive and in danger of infringing. That being said, other scholars have recognized that limitations to the authorial right can open up a space for a relational user to emerge. For example, Abraham Drassinower contends that the author's right needs to be limited through conceptual formulations such as the idea/expression dichotomy (discussed in the previous chapter) in order to support an ongoing right to original expression among the community. The dichotomy supports this right to expression because "ideas can be said to enter the sphere of right, of relations between persons, only [when] expressed";[14] subsequently, under this dichotomy, multiple authors can be encouraged to use existing ideas as part of their politically guaranteed right to expression. Leaving aside the concerns about the usefulness of this dichotomy noted in the previous chapter, this premise at least shows how limitations on authors' rights can construe an alternative user that is presented with a trajectory toward authorship.[15] Along the same lines, the concept of the public domain, consisting of factual material, works out of copyright, and so on, also opens up a space for users to emerge and become prospective authors.[16]

Taking this brief outline of the user in these Commonwealth countries as a starting point, we can identify a highly complicated relational subject that is drawn across piratical and authorial lines. The exceptions granted to copyright infringement construct a user–pirate, as they see every user of copyrighted material as a potential pirate who may be required to defend their use through fair dealing exceptions. Yet a number of limitations to the authorial right offer a contrary space of authorial emergence that recognizes the need to establish equitable relations among current and potential future authors.[17] In addition to these competing trajectories, there is of course an element of the user that is not as prominently articulated in law—the user as a consumer of information and an audience for copyrighted works. Despite the complex legal construction of the user that extends out from acts of consumption to incorporate authorial and piratical capacities, we will soon see how one vision of the user can quickly gain cultural and legal precedence over the others. But before doing so, I will offer a brief history of US copyright law and outline the different system of user exceptions that are found in this jurisdiction.

Fair Use and a Short History of Copyright Law in the United States

US law has a basis in common law due to its history as a colony of Great Britain. Since the US became an independent nation state, however, its legal system—including copyright law—has occasionally diverged from this heritage. The first point of interest is that the United States has a copyright clause embedded in their constitution. It states: "The Congress shall have power ... To promote the progress of science and useful arts, by securing for limited times to authors and inventors the exclusive right to their respective writings and discoveries."[18] The clause does not specify how copyright operates in practice. But in contrast to the gradual development of copyright law in the UK we saw in chapter 1, we could say that from the outset the US had a general idea of the purpose of copyright. It was an instrumentalist bargain between authors (and owners), the US government, and the general public, and the law itself was oriented toward progressing science and culture. This utilitarian approach has been maintained in one way or another throughout the history of US copyright law and is why the law has tended to be viewed in solely economic terms. A copyrighted work is treated as an alienable piece of property instead of an inalienable extension of the author's personality.[19]

When it comes to the user though, the central difference between US jurisdiction and those discussed earlier is the presence of a fair use rather than a fair dealing exception. Fair use has a similar history as it was also gradually developed as a judicial precedent for over a century before it was codified in 1976. The 1841 decision of Justice Joseph Story in *Folsom v. Marsh* is commonly believed to be the origin of this precedent.[20] The case centered around two works that retold the life of George Washington. Plaintiff Jared Sparks had composed a biography of Washington in one volume and then compiled the "correspondence, addresses, messages, and other papers, official and private" from George Washington in eleven other volumes.[21] The work was titled *The Writings of George Washington* and published by Folsom, Wells, and Thurston. It was subsequently alleged that a Reverend Charles Upham had produced a text titled *The Life of Washington*, published by Marsh, Capen and Lyon, which copied 388 pages from Sparks's book consisting of "official letters and documents and ... private letters of Washington."[22]

Justice Story found that *The Life of Washington* was infringing, stating that as Sparks's original work was "mainly founded upon these letters," replicating these letters would affect the "essential value" of *The Writings of George Washington*.[23] As part of this judgment, Story developed some elements of fair abridgment theory, which was already developing in the common law courts of England. Drawing on *Gyles v. Wilcox*, Justice Story noted that for an abridgment to be considered fair, "[t]here must be real, substantial condensation of the materials, and intellectual labor and judgment bestowed thereon; and not merely the facile use of the scissors; or extracts of the essential parts, constituting the chief value of the original work."[24] In addition to this formulation, however, he also noted that when assessing use, it was a question not just of the quantity taken but also of the broader market impact of a competing title. Again referring to common law precedent, Story emphasized that in assessing claims of infringement and use, a judge should also consider "the nature and objects of the selections made, the quantity and value of the materials used, and the degree in which the use may prejudice the sale, or diminish the profits, or supersede the objects, of the original work."[25]

From this origin point, we can see that the notion of fair use was influenced by "a long line of English fair abridgment cases, dating back to the beginning of statutory copyright law in 1710."[26] Still, *Folsom v. Marsh* also offered its own unique lines of legal reasoning, sketching out the principles that would define the fair use doctrine. Justice Story collectively considered the amount copied, the impact on the market of the original work, the purpose of the new work, and whether the work was transformative enough to be considered a new work.[27] These principles were refined procedurally over the following century to the point that a "fair use" judicial test could be confidently applied in the middle of the twentieth century. This test involved a "consideration of (1) the quantity and importance of the portions taken; (2) their relation to the work of which they are a part; (3) the result of their use upon the demand for the copyrighted publication."[28]

Fair use was subsequently codified in the Copyright Act of 1976. Originally, the statute did not define fair use but simply stated that "fair use of a copyrighted work ... is not an infringement."[29] It then set out four factors that courts should use to determine whether a use was fair or not. They would have to consider:

(1) the purpose and character of the use, including whether such use is of a commercial nature or is for nonprofit educational purposes;

(2) the nature of the copyrighted work;

(3) the amount and substantiality of the portion used in relation to the copyrighted work as a whole; and

(4) the effect of the use upon the potential market for or value of the copyrighted work.[30]

The benefit of this long road from judicial precedent to codification was that unlike fair dealing (which offers a specific set of exceptions), the fair use doctrine retained a relatively broad scope. In principle, this meant that US copyright law should have been open to a range of different uses as judges were not restricted by an overly narrow legislative framing. The lack of specificity in the fair use doctrine could be seen as a positive because, according to law, everyone was a potential user.

Despite this promise, fair use was interpreted in an incredibly narrow fashion for much of the late twentieth century.[31] The judiciary did not find favor with commercial uses that could be said to even remotely affect the ability of the original copyright holder to exploit future commercial markets. This was because throughout the period, the fourth fair use factor was viewed as the most important and interpreted relatively strictly,[32] allowing rights holders to hinder a number of "interesting, creative, and culturally significant reuses of their works."[33] Subsequently, the general trend of US case law was to support the rights of existing copyright holders instead of taking a more generous interpretation of the exceptions.[34]

Over the past couple of decades, however, an increasingly liberal interpretation of fair use has been gradually entrenched in copyright jurisprudence.[35] Patricia Aufderheide and Peter Jaszi have mapped out this trajectory in some detail, noting recent decisions that have allowed: filmmakers to use John Lennon's song "Imagine" in a documentary without authorization from Lennon's estate;[36] "documentarians, college teachers of all kinds, film and media studies students, and noncommercial video creators" to break the encryption on commercial DVDs in order to create new works;[37] and a coffee table book on the Grateful Dead to replicate posters without the authorization of the copyright owner.[38] Most remarkably, Aufderheide and Jaszi note that Google's search function can only operate through a broad reading of the fair use doctrine that allows the replication of thumbnail images and small amounts of text.[39] Therefore, the user has undergone something of a change with these recent decisions, with cases such as *Bill Graham Archives* resoundingly departing from existing jurisprudence by accepting secondary uses that may "impact on a copyright holder's potential market."[40]

These developments have produced a user that is perhaps slightly more authorial than the fair dealing user outlined earlier. Whereas the user construed by fair dealing is ascribed a variety of piratical and authorial agencies, recent US jurisprudence has shown that American courts view the user as not just a consumer but a potential producer of noncommercial and commercial texts who should be provided a space from which to engage in authorial activities. That said, there is still a small piratical element that seeps into the concept of fair use because it functions as an affirmative defense to infringement. This requires the user to defend their right to not be positioned as a user–pirate by the law. Lawrence Lessig notes that this often leaves a prospective user with two options, to "either pay a lawyer to defend your fair use rights or pay a lawyer to track down permissions ... a privilege, or perhaps a curse, reserved for the few."[41] Additionally, other scholars believe that these recent developments in fair use jurisprudence have not yet "solved" the problem of fair use. Both Amy Adler and Rebecca Tushnet note that current fair use jurisprudence does not provide clarity around the extent to which artists can copy from previous works and fails to seriously engage with the practices of the contemporary art world or the image as a concept more generally (a discussion that will be taken up in the next chapter).[42]

Regardless, in both the fair dealing and fair use examples discussed, we start to see a relational user emerge, one that resides in an intermediary position and is ascribed authorial and piratical agencies. This relational user is not static, and we see the user move across these various domains of subjectification over time (see figure 2.1). The story of fair use jurisprudence moves the user away from the author and associates the subject with the pirate before returning the user to its authorial alignment. Moreover, the introduction of fair use and fair dealing signaled a marked change from the pre-1911 approach to the user, when use was treated in a comparatively ad-hoc manner. These changes have been as much cultural as legal, however, and the

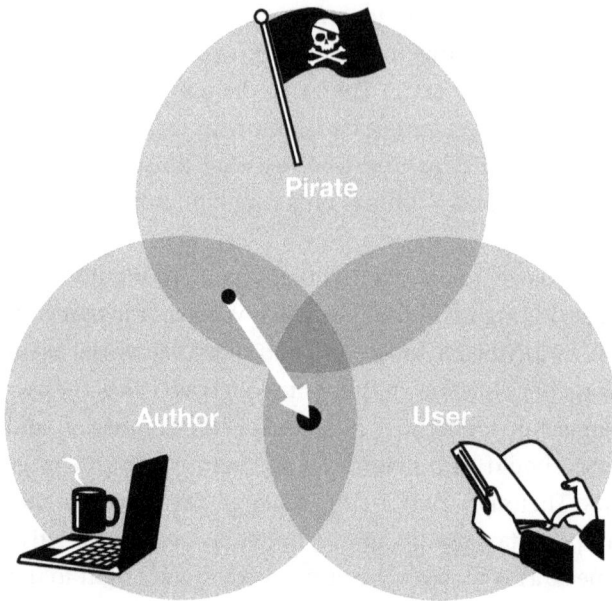

Figure 2.1
A number of scholars have argued that recent judicial interpretations of the United States' fair use doctrine increasingly recognize an authorial user.

rest of the chapter will be dedicated to an examination of the associated cultural developments that have influenced how different jurisdictions have interpreted the user. The development of media during the twentieth and twenty-first centuries maps onto the changing status of the user and law and shows how these legal modes of subjectification were deployed in the context of broader changes in the contemporary media environment.

New Media and the New User

We can identify a clear transition occurring around the user through the study of the emergence and eventual mainstream adoption of radio and television. Although the radio originally allowed users to function as active creators and consumers, through a combination of regulation, product design, and the normalization of particular types of media use, individuals were gradually positioned as consumers of copyrighted works. We can then see how a range of "innovations and adaptations" around broadcast television in the latter half of the twentieth century (most notably the VCR, or videocassette recorder) challenged this passivity and allowed individuals to gain control and agency over their broadcast signal.[43] In this section, copyright law recedes into the background as we attend to regulation, media use, and technological change. This perspective shows how other actors such as distribution mediums can contribute to the construction of subjects within the relational triad.[44]

The story of the radio is an interesting one because the medium itself directly embodies the shift from an active to passive user. The radio was initially a user-led invention, with hobbyists developing the medium through experimentation. Since the spectrum was not yet properly regulated, amateurs dotted the airwaves using homemade sets to exchange messages with one another and even broadcast to the world at large. Thanks to this "grassroots subculture" in the United States, radio stations "grew rapidly ... from about 150 in 1905 ... to more than 10,000 in 1914."[45] These amateurs were required to stop their experimentation once World War I began, in order to keep the spectrum free for military operations, but once the war ended, they returned to their previous work.[46] In these early days of radio anybody with sufficient know-how could not only produce and distribute content en masse, but also use the medium to send and receive messages (much like the Internet today).

These experiments led to a demand for both regulation of the spectrum and a greater professionalization of radio because of amateur interference with military and commercial communications.[47] The US government responded with the Radio Act of 1912, which declared that no one could "operate a radio without a license from the Department of Commerce and Labor."[48] Many amateurs simply ignored this new rule, however, and the department did not have sufficient funds to enforce the law. Radio became an increasingly popular medium during the 1920s, with a number of new commercial stations appearing on the band (often amateurs commercializing their hobby), but despite these new faces, for the majority of the population "private broadcasting" was still "the activity people most closely associated with radio."[49] Indeed, in 1920 there were still "15 times more amateur stations than all other types combined," as well as a number of nonprofit stations including churches, municipalities, and educational institutions.[50]

This led to something of a crisis with the spectrum becoming increasingly chaotic as more stations came on the air. The first fundamental change to combat the chaos was the introduction of a new regulation in 1921 that declared that amateur licenses could not broadcast music, weather, news, speeches, or anything else of interest that would compete with commercial stations.[51] This effectively neutered the amateur use of radio for entertainment and "radio became a more passive medium."[52] The next major step in the regulation of radio was the Radio Act of 1927.[53] This piece of legislation established the Federal Radio Commission, a body that would be solely responsible for the issuing of licenses. Importantly, though, this reform also dramatically limited the number of licenses on offer, eliminating nonprofit stations and favoring broad-based stations over the narrow-interest broadcasting that had defined the 1920s.[54] The act signaled the victory of commercial broadcasting over amateur experimentation and noncommercial use, with the radio now a medium to "listen to" rather than engage with.

In a few decades, radio again changed dramatically. With radio now a largely corporate environment, it was able to seamlessly connect to cultural industries like the music industry and over time a series of professional conventions emerged, such as the formation of music industry bodies to negotiate and approve the licensing of copyrighted music for public performance.[55] These professional networks allowed for a simple negotiation of complex legal frameworks like copyright law among a few key stakeholders. This reconfigured the law itself, turning it into a form of specialized knowledge and a text to be mobilized by content industries and their respective legal teams, and the everyday person as "user" was nowhere to be seen.[56] Instead, the user of musical recordings was now interpellated as the radio company, who in turn functioned as a key marketing arm for the music industry.

Television interpellated the user subject in a different manner. The first obvious point to make is that television did not have a two-way capacity like radio, so from the outset the majority of individuals were framed in a passive position as television "audiences." This passivity was also entrenched through regulation. For example, in Britain both radio and television audiences were actively regulated by the state in the form of a license fee. Essentially, this required an individual to pay for an annual license in order to own and operate a radio or television. In the early years of radio, licenses were 10 shillings, and when television was first introduced the service was incorporated into this initial fee. As the medium developed, these charges were separated and from June 1, 1946, a household could purchase a radio license for one pound or a combined radio and television license for two pounds.[57] The practice has continued to this day and British people currently have to pay an annual fee of £147.[58] Indeed, the Communications Act 2003 states that a "television receiver must not be installed or used unless the installation and use of the receiver is authorised by a licence."[59]

This license fee is administered to support the activities of the BBC (British Broadcasting Corporation). Until September 1955, though, there were no competing commercial channels on the air in Britain.[60] Therefore, for a number of years, households had to pay a license fee to directly support the sole broadcaster and distributor of television content in the country. This is of interest because it underlines just how closely the state and distributors of mass media interact to support particular cultural markets and define the circumstances through which audiences can gain access. During the era of broadcast media, the ability to access and consume copyrighted content was mostly demarcated by this sort of regulation just as much as, if not more so than, copyright law. And once again, much like radio, the decision occurred in conversation between the regulators and distributors of media, with the general public interpellated as consumers who were supposedly happy with a limited menu of content.

These histories of television and radio underscore two important points about how the user was understood for much of the twentieth century. First, the brief outline of US radio's gradual commercialization and the administration of the license fee in Britain show: how active users can be controlled to limit their productive capacities on a medium (radio); how industry structures can be organized around the transfer of copyright (radio); and how regulatory barriers are created around content access (television). These cases demonstrate the importance of regulation and industry practice in shaping markets around copyrighted goods and defining terms of access to copyrighted content; in turn, this process contributes to the interpellation of the user. The development of this subject occurred not only through the courts but also through the regulatory and technical organization of broadcast media, which carried particular imaginaries about what people do with technology and the extent to which individuals could (or should) engage with copyrighted content. Second, these narratives highlight the important role of intermediaries as key mediators of copyright law. Socio-legal discourses tend to speak of the author, user, and pirate with distributors and other intermediaries as absent subjects. Yet, as we have seen, industry bodies and distributors set the terms around accessing, licensing, and broadcasting copyrighted content.

This cultural and legal framing of the general public as passive users began to be challenged in the latter half of the century, thanks to a range of tele-visual "innovations and adaptations" that extended the frontiers of domes-tic television.[61] By the 1960s, transistor televisions allowed TVs to become "smaller, cheaper, lighter and portable," introducing an era of mobile tele-vision well before the age of the laptop and smartphone.[62] By the 1970s and 1980s, domestic audiences were starting to gain direct control over content and the medium itself thanks to programmable televisions, the introduc-tion of remote controls, and, perhaps most dramatically, the programmable VCR.

Although it is important not to over-emphasize the impact of techno-logical development during the 1980s, articles about the VCR emphasized the growth in audience agency, and there was a real sense of being liberated from the constraints of broadcast media. *Newsweek* positioned the VCR at the heart of a broader media revolution and suggested that the technol-ogy gave "power to the people."[63] The magazine argued that the technol-ogy directly challenged the "tyranny" of TV, not just through the ability to time shift or watch restricted content but also by freeing the audience from the strictures of television programming (see figure 2.2). As one VCR user Barbara Riley explained, "We used to watch a lot of shows to get to a show ... Now we just set the timer ... We're watching less TV."[64] The emer-gence of "mom-and-pop retail stores" providing a wealth of rental and purchasing options to consumers also contributed to this new consumer freedom that allowed the general public to choose how they consumed copyrighted content.[65]

This sense of agency was incorporated into media and cultural stud-ies research conducted during this period. No longer were the media and popular culture functioning as an all-encompassing ideological bludgeon; instead, audiences were being repositioned as active participants in the industrial structures of mass media. Scholars critiqued previous research that situated audiences as passive consumers of media and instead argued that viewers were not only able to retain a critical distance from popular culture and the media, but were able to produce their own divergent mean-ings and associations. This work was receptive to the changing perceptions

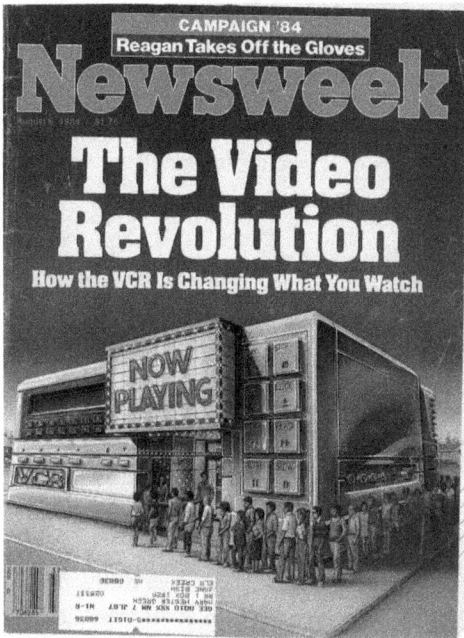

Figure 2.2
The VCR was part of a broader change around how individuals related to copyrighted media.

around the relationship between consumers and media and contextualized this moment of technological change.

To take one example, Ien Ang outlined the highly material ways the VCR was providing audiences with a new feeling of agency. The VCR was especially popular among marginalized groups such as migrants and other audiences (such as viewers of hardcore pornography) who had been poorly served by "centralist, modernist television."[66] The ability to access filmed content (either ethnic or obscene) that would never appear on broadcast television, record television programs at home to watch at a later date, and even rent movies on a whim created a sense of agency, with people freed from the strictures of the television program guide. Broad swathes of media users no longer identified with the passive mass audience imagined by the broadcast media industry, and that image slowly gave way to an "active" individual user that retained both agency and choice.[67]

These broader cultural changes eventually made their mark on copyright law in the 1984 *Betamax* Supreme Court decision that ruled that VCRs should be deemed legal due to "their potential for productive non-infringing and fair uses."[68] The decision acknowledged that copyright law could not restrict innovative technology that could potentially be used for non-infringing purposes. Indeed, this was the eventual story of the VCR as it was largely employed for "consumptive use."[69] Lucas Hilderbrand argues that in terms of a strict reading of the fair use exceptions, the dissenting opinion in *Betamax* may actually be the more legally persuasive.[70] Therefore, the fact that the majority opinion took such a generous interpretation of fair use suggests that these emerging cultures of use, and indeed VCRs themselves, influenced the judgment of the Supreme Court.[71]

The historical trajectory outlined in this section shows how the user was understood culturally throughout the twentieth century. Though audiences were extremely active during the early years of radio, by the middle of the century, regulatory developments around radio and television broadcasting had solidified a notion of the user as a consumer of copyrighted material. It was not until the emergence of technologies like the VCR that the general public regained a sense of agency around media consumption. In addition, industry arrangements saw intermediaries like radio stations claim the title of "copyright user," at times leaving the public out of the copyright picture entirely. This narrative demonstrates that a range of broader factors shape and contextualize the operation of copyright law. As we gradually move toward the twenty-first century, an examination of Web 2.0 and the emergence of the creative user reveal the ever-increasing cultural agency of the user.

The Birth of the Creative User

During the early years of the World Wide Web, an individual's online experience was largely typified by a page-oriented infrastructure with web pages connected to each other through hyperlinks and web rings. Following the turn of the millennium, however, a number of innovative websites were progressively launched that presented new opportunities for working, engaging, and otherwise interacting online. The crowd-sourced encyclopedia *Wikipedia* (2001), social networks MySpace (2004) and Facebook (2004), and video-sharing website YouTube (2005) (to pick just a few well-known examples) emphasized creating content, sharing experiences, contributing to a broader knowledge base, and strengthening social ties. These websites were incredibly popular and for many scholars, the emergent online practices they encouraged not only created new trajectories of knowledge and sociality but also represented a radical shift in discussions around creativity and use.[72]

Yochai Benkler has argued that these changes allowed users to take a more active role in cultural production and has suggested that they allowed the general public to become "a potential speaker, as opposed to simply a listener and a voter," clearly suggesting that there had been a shift from passive use and consumption to active engagement.[73] Clay Shirky has also addressed this moment of transition, noting that these practices challenged traditional institutionally sanctioned forms of content creation and information dissemination such as journalism.[74] Henry Jenkins contributed to this approach through his work on fandom.[75] He has outlined a trajectory of increased inclusion in which fandom went from the margins of culture to playing a central role in cultural production, with a new participatory culture allowing media producers and consumers to "interact with each other according to a new set of rules."[76]

Strangely enough, *Time* magazine was the most radical interpreter of this moment in the cultural zeitgeist. In 2006, the magazine famously announced that "You" was the person of the year, collectively awarding this hallowed title—one that had previously been held by the likes of Winston Churchill, Adolf Hitler, and Mahatma Gandhi—to all of us. Inside the magazine, emblazoned with a reflective mirror cover, was a paean to the Web 2.0 era:

We made Facebook profiles and Second Life avatars and reviewed books at Amazon and recorded podcasts. We blogged about our candidates losing and wrote songs about getting dumped. We camcordered bombing runs and built open-source software.[77]

The profile embodied the utopian ideology of Silicon Valley but managed to take it one step further, with Lev Grossman arguing that although "Silicon Valley consultants call it Web 2.0, as if it were a new version of some old software ... it's really a revolution."[78] Tying together the various strands of cultural ephemera that referred to this particular moment in the zeitgeist, the *Time* magazine profile represented the decisive end point of these discussions: the identification of a new cultural subject, the creative user.

One could easily challenge the novelty of Web 2.0 by noting that since the birth of the Internet, users have been participating actively in communities through bulletin boards, web forums, and mailing lists. Furthermore, as Jean Burgess has pointed out, Apple mobilized a similar discourse of agency and creativity around the user subject as early as the mid-1980s.[79] The real change may not have been the practices themselves, which have a long historical provenance, but rather their scale. Once a niche activity limited to early adopters of technology, this notion of individuals producing cultural goods or remixing existing copyrighted works became a form of everyday media practice. It is important to recognize the historical specificity of the Web 2.0 discourse, while also acknowledging that contemporary cultural understandings of the creative user can build on this history and enable the rapid emergence of "putatively new forms of media subjectivity."[80]

The new cultural discourse that has emerged around the subject of the user has contributed to a change in how the user is understood, with this subject now positioned as a creator rather than a consumer. Moreover, this cultural narrative has developed an interstitial and wholly relational subject in the form of the creative user—an actor who can straddle the activities of use and creation concurrently. This is particularly noticeable in the various neologisms that have been coined to describe these practices, from "consumer cocreation"[81] and "user-generated content" (UGC) to "produsage."[82] Although there are minor differences between the definitions, all of these terms attempt to speak to the uncertain phenomenon of large-scale amateur creativity and, in turn, attempt to link the user to some sort of creative authorial capacity (see figure 2.3).

This new cultural narrative around the user, reacting to developments occurring from the early 1990s onward, did not, however, spark a transformation in fair use jurisprudence.[83] But the growth in judicial tolerance for fair use actions did dovetail neatly with the emergence of amateur media online. Indeed, one can see the importance of this cultural and legal reconceptualization of the user in the infamous *Warner Bros. Entertainment, Inc. and J. K. Rowling v. RDR Books* case, which saw Warner Brothers and J. K. Rowling sue Steven Vander Ark, the operator of the website *The Harry Potter Lexicon*, for copyright infringement.[84] The *Lexicon* functioned as a "reference guide for fellow fans to navigate [the] series" and a fictional universe that was expanding and becoming increasingly complex.[85] Although Vander Ark ultimately lost the case because "[the Lexicon's] actual use of the copyrighted works [was] not consistently transformative," Judge

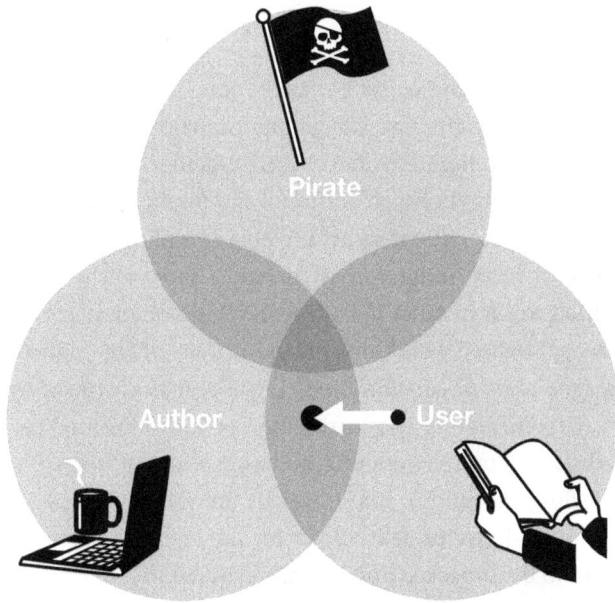

Figure 2.3
Commentary around Web 2.0 ascribed an increasing amount of creative agency to the user.

Robert P. Patterson recognized that the *Lexicon* was a broadly transformative work.[86] This ruling not only "vindicated" the transformative use principle but also stood as evidence of a general judicial openness to the new forms of creativity present online.[87]

In contrast, the system of fair dealing has been directly influenced by the new user narrative, and a number of jurisdictions have introduced new fair dealing exceptions in an effort to respond to recent changes around use and creation, from a parody and satire exception in Australia[88] to a user-generated content exception in Canada.[89] I will examine these changes in chapter 4 as well as explore the tensions and complications that have emerged around this notion of the creative user, particularly with respect to actors who sit somewhere between the amateur/professional divide. At this stage, however, it is enough to note that across the jurisdictions this book considers, law and culture have moved in relatively concurrent trajectories away from a twentieth-century vision of the user–pirate and have started to embrace the notion of a creative, authorial user.

Conclusion

In this chapter, we have seen how copyright exceptions and limitations, principles that underpin copyright law like public domain, and broader cultural and technological developments have come together to produce a relational user, one that oscillates between authorial and piratical capacities. At a general level, this gives us some insight into this complicated subject that can fulfill a range of roles under copyright law: a proto-author that needs to use copyrighted works in order to engage in the production of new works, a consumer of copyrighted works, or a proto-infringer of copyrighted works either through methods of consumption (i.e., illegally downloading a film), distribution (i.e., circulating the infringing film), or creation (i.e., using more than a substantial part of a copyrighted work without protection from a fair use/fair dealing exception). It also signals the value of engaging in relational analyses of subjectivity, as it allows for the complex constitution of the user subject to be unpacked and examined in detail.

Additionally, an overarching socio-legal narrative has also been identified that has seen technological, cultural, and legal developments combine to move the user away from the author and toward the pirate before slowly restoring the user–author around the turn of the twenty-first century. This narrative contributes to our reading of the relational user as it shows how technological developments, copyright jurisprudence, and copyright policy can directly affect the constitution of the divergent strands of the user. We can identify a user–pirate who is positioned almost in opposition to the author and commercial activity amid the narrowing of the fair use doctrine in the twentieth century, and a shift toward an authorial user following the more generous judicial interpretations that occurred from the early 1990s onward. These changes in jurisprudence developed in tandem with cultural changes that at times offered their own vision of the user, as seen with the emergence of the "creative user" subject amid the Web 2.0 hype in the mid-2000s.

Subsequently, this analysis of social and cultural trends, technological developments, and media regulation also outlines the critical role each factor plays in the process of subjectification. We have seen how these cultural shifts have contributed to broader ideologies regarding the capacities and agencies of particular subjects, and how they have intertwined with jurisprudential trends to coalesce around an interpretation of the relational user as a proto-author, consumer, or possible pirate. Moreover, although the presence of a specific ideological approach to the user has often been quite clear in the legal arena, these cultural developments have signaled how ideologies and subjectification can occur through more complex interactions between law, technology, and the use of media in our everyday lives. The following chapter continues to examine these interactions between culture, law, and subjectivity and extends this analysis of the tensions between use, authorship, and infringement through a study of the relational author in the context of unauthorized sampling and the contemporary art industry.

3   Creative Practice and Relational Authorship

"Blurred Lines" was one of the most popular songs of 2013. It spent a record sixteen consecutive weeks as the number one single on the Billboard Hot R&B/Hip-Hop Songs chart and "sold 5.8 million downloads in its [first] 26 weeks of release."[1] A staple of dance floors and Spotify playlists, the song made a significant amount of money for its writing team, Robin Thicke, Pharrell Williams, and T.I. (a.k.a. Clifford Joseph Harris, Jr.). But the song rested on a catchy groove and Fender Rhodes piano line that sounded like Marvin Gaye's track "Got to Give It Up" to many knowledgeable listeners.[2] Members of Gaye's estate also noted this similarity and threatened to sue the writing team for copyright infringement unless they were paid a monetary settlement. The estate rejected a six-figure settlement offer and so Thicke and Williams "filed a pre-emptive lawsuit" in August 2013, seeking a ruling stating that the song did not infringe Gaye's copyright.[3] After a two-week trial, a jury decided that "Blurred Lines" did in fact infringe copyright and the writing team was ordered to pay Gaye's estate US$7.4 million.[4] The case is currently being appealed.[5]

From the release of "Blurred Lines" through the subsequent trial, acts of authorship, use, and theft emerged at different times. The song's release saw Thicke and Williams lauded as creative individuals who firmly deserved the authorial mantle. The Gaye estate's legal challenge then repositioned these two artists as pirates who had stolen key elements from Gaye's composition. In response, the "Blurred Lines" writers mounted a defense of their authorship, but they also admitted during the trial that they may have "used" musical elements that were similar to those in Gaye's song. Furthermore, Thicke revealed that despite being named as an author, his creative contribution was incredibly minimal as he was high on Vicodin for much of the writing process. This reminds us that authorship is not just a creative process but also one embedded in a series of broader social, cultural, economic, and biological relations. Thicke may have been granted authorship in spite of his limited creative contribution to ensure the receipt of potential royalties following his central role as a vocalist on the track.[6]

This brief tale offers an excellent example of the complex ways in which legal subjectivity and authorship are constituted and understood in contemporary copyright law. Thicke and Williams could be considered authors or, as the jury eventually decided, pirates. Indeed, Marvin Gaye's own authorship rests on a long history of collaboratively authored rhythm and blues, jazz, and gospel music, which influenced and informed his stylistic choices as an artist.[7] Existing theories of relational authorship view culture as the output of a communicative process that is collaboratively developed through creative practice and argue that authors must be allowed to use existing cultural resources to build on what has gone before them.[8] Both the "Blurred Lines" case and the initial investigation of authorship, use, and infringement undertaken in previous chapters has shown, however, that copyright doctrine manages relational authorship in a varied fashion, allowing some uses and rejecting others.

This chapter continues to explore relational authorship by turning to the music and art industries to examine how they interact with these doctrinal boundaries. These two industries recruit or ignore copyright doctrine strategically in order to set up their own systems of authorship and attribution, and we will see how these actions affect how use and infringement are construed. This study of industry also provides an example of how subjectivity in copyright is constructed in relation to the formation and maintenance of markets. Furthermore, institutions and industries can produce relationality (and the relational author) in different ways, which raises questions about the effectiveness of copyright law as a regulatory system. It also highlights how alternative industry-based interpretations of the relational triad can be equally valid in the view of many practitioners. Attribution also appears as a subset of this chapter's focus on the authorial claim. Linking a name to either an original or derivative copyrighted work is central to the maintenance of an author's presence and authority. It establishes a work's lineage by allowing someone to claim ownership over a work or to draw reference to a work's history by marking out the creative heritage the work has drawn on. It also helps to sustain the reputation of the individual and the work over time, building cultural and economic value that allows both to function as assets in the wider marketplace. Lastly, copyright and particular creative industries have a number of idiosyncratic ways of managing the relationships between different copyrighted works, and various ways of interpellating the individuals that contribute to their creation. Examining these contestations and divergent approaches will also form part of this chapter's analysis.

The chapter begins by providing an overview of music licensing, tracking the shift from composers and artists freely quoting and borrowing other people's music to the current system of licensing samples. I will then briefly discuss a recent case in Australia that was similar to the "Blurred Lines" decision previously discussed—*Larrikin Music Publishing Pty Ltd. v. EMI Songs Australia Pty Ltd.*—before engaging in a broader analysis of music licensing. I outline how this system (in conjunction with copyright law) attempts to regulate subject positions, provide a formal method of attribution, and function as a source of income. We then turn to the world of art and consider the fact that authorship, attribution, and use function in an entirely different manner in this world. In the art industry, not all who create are authors and not all who copy are pirates, revealing the extent to which industry practice shapes subjectivities.

Borrowing, Quoting, and Sampling: Music Licensing and Regulating Use

The Western musical tradition is built on borrowing and sampling. Brahms, for example, was a great sampler. His Third Symphony borrows from a variation of the main theme of Robert Schumann's Symphony no. 3 ("Rhenish"). Beethoven was an expert borrower, studying Mozart's A major string quartet so intently that his own A major quartet mimicked Mozart's structure.[9] We might even consider the work of Rachmaninoff, whose "Rhapsody on a Theme of Paganini" leant heavily on Paganini's 24th Caprice in A minor. This borrowing and development of melodies and compositional structures was not only unremarkable but seen as integral to the general process of creativity. Indeed, the importance of quotation to classical composition is underlined by the popularity of an unattributed saying (erroneously attributed to Igor Stravinsky) that states: "good composers borrow and great ones steal."

Jazz has similarly rested on a history of borrowing and sampling. Groups and musicians regularly worked off charts that outlined the melody and chord changes of standard songs (or "standards"). Recognizable melodies were quoted during instrument solos, existing chordal structures of songs formed the basis for new works, various stylistic trends (such as bebop or swing) informed the development of solos, and bands regularly covered popular songs (and of course, still do to this day). These practices were embedded in the genre's cultural roots, which emerged from African American spirituals centered on an oral tradition and a "tradition of quoting."[10] In terms of formal law these practices interact with copyright in a number of strange ways. For example, it has been argued that working out authorial origin (and, more important, where economic benefits should flow) in this iterative and constantly changing genre is incredibly difficult.[11] Despite these problems, however, the genre only very occasionally ran into copyright issues during the twentieth century.[12]

This very brief account of the Western musical tradition's tolerance of sampling, borrowing, and quotation is offered because it provides context for this section's discussion of hip-hop and the sampler. The story starts with hip-hop artists using the sampler to continue this practice and ends with a system of music licensing placing new economic restrictions on this sort of quoting and repurposing and effectively changing the course of sampling forever.

In the early 1980s two new musical technologies came onto the high-end consumer market: the sampler, which could record and play back sound samples at the press of a button, and the drum machine, which could produce digital or analog beats (and occasionally also sample sounds). Noted hip-hop artists like Grandmaster Flash had already been collecting records and physically sampling material through the novel use of two turntables and a mixer, offering a backbeat for rappers to freestyle over. These new instruments became particularly useful as hip-hop made the transition from a niche genre to a chart-topping phenomenon and major artists started to enter the recording studio. Samplers provided artists with an effective and powerful way of producing a collage of sounds to support MCs and were increasingly prominent in popular music as the technology became cheaper and cheaper across the decade.[13]

Some of the most notable uses of this technology were by Public Enemy, who used samples liberally in their album *It Takes a Nation of Millions to Hold Us Back*. Rather than simply rapping over "the entire rhythm of a song," Public Enemy took "a horn hit here, a guitar riff there ... a little speech, a kicking snare from somewhere else" and worked them into a phenomenal collage.[14] Dorian Lynskey explains the density of this sampling and recontextualization, noting that you can

unpack a whole history of black music and resistance from the samples and quotes of James Brown, Gil Scott-Heron, John Coltrane, Bob Marley, Malcolm X, Jesse Jackson and so on, but you can also hear it assimilating white culture in the form of, among others, Queen, Slayer and James Dean's Rebel Without a Cause.[15]

Other notable examples of this sample-heavy period in hip-hop are De La Soul's *3 Feet High and Rising* and the Beastie Boys' *Paul's Boutique*, where "95 percent of the sounds on [the] record came from sampled sources."[16]

This stellar creativity was not so evident in other work of this period. Many rap artists just "looped the hook of an earlier song" and, as Kembrew McLeod and Peter DiCola explain, this often led the sampled artists (who could clearly identify their songs) to contact the sampler directly and simply demand payment, promising not to follow through with a drawn-out legal action.[17] This practice was so effective that publishers and artists started to view it as a useful method of income generation, leading to something of a sampling arms race. Record label representatives of sampled artists were requiring more and more of their copyrighted material to be licensed and the representatives of artists who sampled were becoming fearful of the increasing costs involved as well as the potential legal ramifications. Whereas the albums discussed above needed to clear only a handful of samples, by the early 1990s artists were required to not only clear more samples but weigh "the legal risk of *not* clearing samples," as the music industry (and its lawyers) had started to pay attention to this burgeoning practice.[18]

This litigious climate was also informed by contemporaneous developments in copyright jurisprudence. The very first case to consider sampling in the US was *Grand Upright Music Ltd. v. Warner Bros. Records, Inc.* and it is notable that the final judgment opened with the biblical admonishment "Thou Shalt Not Steal."[19] Biz Markie had asked to use a sample of the song "Alone Again (Naturally)," written by Gilbert O'Sullivan, for his own song "Alone Again." Markie's request was refused and he decided to use the sample anyway. This act naturally led to a lawsuit. In the case, Judge Kevin Duffy found that the use of the sample was infringing, issued a preliminary injunction, and suggested that Markie should also be referred for criminal prosecution. According to this interpretation, copyright infringement

would be the automatic result "once the plaintiff prove[d] copyright ownership and unauthorized sampling."[20] At around the same time, the Supreme
Court judgment in *Campbell v. Acuff-Rose Music* authorized sampling without licensing in cases where the intent was clearly to parody. But this was
a risky legal interpretation to rely on as a general foundation for a right to
borrow or quote music.[21] 2 Live Crew's manager had originally sought a
license from the publisher of Roy Orbison's song "Oh, Pretty Woman," and
the parody argument only arose in the context of the license being denied.
These facts were directly relevant to fair use assessment. Some have suggested that the general tenor of these decisions spooked the music industry
to the extent that from this point on, major label artists were required to
license *any* samples they wanted to use prior to publication.[22]

Johnson Okpaluba has advanced an alternative interpretation to the aforementioned narrative, which stands as the dominant scholarly (and popular)
account of the birth and subsequent "death" of sampling. He contends that
the absence of court cases during the height of digital sampling in the mid-
to late 1980s did not mean that there was an absence of official regulation.
Okpaluba notes that "some major publishers and record companies developed detailed internal guidelines to promote and regulate the licensing of
samples," which showed that the industry recognized the value of sampling
and did not immediately set out to stifle it.[23] He goes on to note that following the *Grand Upright* decision, sample-heavy records were still released and
the recording industry also engaged in a range of innovative partnerships to
support particular sampling practices.[24] These corrections present a series of
welcome complications to the popular narrative about sampling and the law
that is often told.

Nevertheless, it is generally accepted that sampling became both more
formalized and more difficult to navigate following *Grand Upright*, with a
set of business practices established that are still in operation today. They
require the party requesting the sample to work with a sampling clearance
house to source and enter into negotiations with the relevant copyright
holder, which is often a record company as opposed to an individual.
Through these negotiations the parties come up with a final amount for
using the sample that takes into account a number of factors, such as how
much of the song is used, the importance of the section used, and the commercial potential of the new song.[25] The payment also has to take into
account the clearing house's labor costs for finding the original copyright

holder and, further, involve paying for two licenses: one for the recording itself and one for the musical composition.[26] Subsequently, sampling has become an increasingly expensive business. Whereas artists often had to pay only a relatively small amount of money to sample tracks in the mid-1980s (indeed, if they were challenged at all by copyright holders), nowadays it costs a significant amount of money to sample even a small amount of copyrighted content and the practice has (arguably) declined accordingly.

Despite this system, legal disputes have still emerged around sampling and US law has continued to treat the practice poorly. Indeed, *Bridgeport Music, Inc. v. Dimension Films*—the most noteworthy case in recent years—has only further entrenched the "second-class" status of sampling in US copyright law.[27] The case stemmed from "Bridgeport Music, Southfield Music, Westbound Records, and Nine Records" alleging "476 counts of copyright infringement against nearly 800 defendants for the unauthorized use of samples in hip-hop music recordings," which were subsequently split into individual actions.[28] One action involved Bridgeport and Westbound claiming that rap group N.W.A. had infringed their copyright by sampling a part of "Get Off Your Ass and Jam" by George Clinton, Jr. and the Funkadelics in their song "100 Miles." N.W.A. had sampled two seconds of a three-note guitar riff that opened "Get Off Your Ass and Jam," lowered the pitch of the sample, and looped it, extending the two-second sample for sixteen beats.[29]

The case was initially dismissed. Although chords cannot be copyrighted, the District Court for the Middle District of Tennessee focused on "the use of and the aural effect produced by the way the notes and the chord are played, especially here where copying of the sound recording is at issue" and found that under this interpretation, the sample was "original and creative and therefore entitled to copyright protection."[30] But the court ultimately found that the sample was *de minimis*—that is, the sample taken was so small that the law should not intervene—and so found for the defendant. Bridgeport went on to win on appeal, however, as the Court of Appeals for the Sixth Circuit offered a stricter interpretation of sampling, stating that this use was infringing and people should "[g]et a license or don't sample."[31]

The trajectory described in this section shows a clear mismatch between the approach to quoting and sampling by artists and by law. There has really been little change artistically from the quoting and sampling practices of classical composers and jazz musicians to hip-hop artists today. The only meaningful differences are the technologies used to engage in the practice and the development of copyright law alongside the emergence of the commercial sound recording industry from the early twentieth century onward. These developments have caused digital sampling to come under fire with the industry treating any reuse of copyrighted material (no matter how small) that is not licensed or does not fall under fair use or fair dealing exceptions as potentially infringing. To cope with this, artists now hire session musicians to play the sample as this means they only have to compensate the songwriter by paying a publishing fee and not a licensing fee and publishing fee.[32] The strange contortions that copyright law forces artists to engage in affects how authorship and subjectivity is understood in relation to musical works, a discussion I take up in the next section.

Music Licensing and Substantial Similarity: Implications for Authorship

The music industry's licensing structure is ostensibly about making sure that the hard work of artists is recognized through attribution and that their work is protected from being liberally used in derivative works. But in a business dominated by multinationals, the industry is essentially organized in this fashion to pursue profits on a global scale in the most efficient manner possible. Therefore, intermediaries make fundamental decisions affecting how relationships between authors, users, and infringers are organized in the music industry.[33] As a number of detailed and carefully researched histories make clear, licensing arguably emerged in order to exploit a growing income stream and now even the smallest "contribution" needs to be paid for.[34] This strict approach still presents us with an acknowledgment of the relational element of creativity that is embedded in the foundations of copyright law, but artists often have little direct control, and recognition of other artists occurs within strictly defined limits. The artistic relationship with other artists is transformed (or reduced) into an economic transaction, usually organized by the artist's label.

Nonetheless, the music licensing system acknowledges relationality. It recognizes that an artist can build on the work of artists that have come before them and obviously allows direct quotation to occur. There is explicit accommodation of the reality that in order to produce new works, many artists are also users. If a licensing fee is not paid for the use, however, the individual becomes an author (by creating a new work), a user (by using an existing work), and a potential pirate (by using this work in an unauthorized fashion) if the copyright holder chooses to enforce their rights. The practice of licensing forces the relationality that is inherent in authorship to be structured in terms of economic power, which leads to significant inequities. It also reveals a number of problems when the opaque nature of the creative process intersects with subjectivity.

As we have already seen, licensing a section of a song costs a significant amount of money and is embedded in a complicated system of approval that relies on specialized industry knowledge. Therefore, whereas in theory anyone can license a sample, it is really a privilege limited to a handful of major artists and their representatives. The structure of the industry opens a significant number of would-be samplers to the often unwanted subject position of the pirate. This system acknowledges relationality only in order to reduce it to an economic transaction. It also presumes that a license is always necessary regardless of the sample size. This injects an often unwarranted authorial lineage and thus maintains one or more spectral figures in the new, usually transformative work. Although in one sense this can be read as a generous move that recognizes those that have gone before, it runs counter to the Western musical tradition in which this sort of quoting and borrowing was seen as standard and not always worthy of acknowledgment. The maintenance of this authorial lineage also means that new works require an imprimatur from existing artists—or as is more often the case, their intermediaries, the record labels—through a drawn-out process that does not necessarily support future cultural production.

This system of licensing is also incredibly ambitious. It attempts to organize, structure, and delineate a series of relationships between subjects with the goal of articulating the moral and economic debts incurred throughout the creative process. This is largely because the licensing system—like many other industrial and economic systems that deal with copyright—operates under a fundamental misreading of copyright law. As Carys Craig explains, both industry and many courts wrongly objectify the rights that copyright bestows, drawing parallels between these rights' "physical and private property concepts" and producing a "metaphorical 'thingification' of expression."[35] In short, the licensing system attempts to produce clearly defined subjects and authorial lineages so that artifacts can be owned and traded.

If we consider how subjectivity can be constructed with reference to the sampling system, however, we see an incredibly complicated series of inter-relations between authorship, use, and infringement that challenges its simplification of the creative process. This becomes particularly noticeable because music

is inherently relational in its construction: the harmonic meaning of a particular note or series of notes depends on the context of those notes. In addition, music is typically related in some way to performance, which distinguishes it from other types of cultural production, such as literature. Music is often less representational than literature, which also strains the relationship between copyright and music.[36]

Therefore, although sampling has produced a legally accepted industrial econ-omy, it does not then follow that this licensing system logically addresses the numerous complexities and contradictions apparent in the interactions between creativity, music, and copyright, which are articulated in the follow-ing case.

The difficult issues of influence, inspiration, and authorship came to the fore in the notable Australian case *Larrikin Music Publishing Pty Ltd. v. EMI Songs Australia Pty Ltd.*[37] Like the "Blurred Lines" action that opened this chapter, this case saw a court called on to make judgments about the similarities between two musical works. *Larrikin Music* offers a much more productive space for analysis, however, due to the strange collection of fac-tors involved: a copyrighted folk song dear to many Australians' hearts, a deceased music teacher associated with an Australian girl scout movement, a legendary 1980s rock band—Men at Work—who wrote a hit that defined the nation, and a small (but litigious) Australian record company.

The facts of the case are as follows. Men at Work released the song "Down Under" in 1981 and the song quickly found success internationally as it appeared to capture something unique about the Australian experience for many people. The song became a mainstay in the Australian cultural imagination and so it was no surprise that it was featured one night in 2007 as a topic on the popular music trivia show *Spicks and Specks*. The host asked participants to "name the Australian nursery rhyme this riff has been based on" before playing a segment of the flute solo from "Down Under."[38] The answer was "Kookaburra Sits in the Old Gum Tree," a popular children's song and another cultural touchstone for many Australians. Most people were unaware not only that what felt like a national folk song was still in copyright but that the copyright holder was still active. Marion Sinclair

composed the tune in 1932 and following her death in 1988, the execu-
tor of her estate, the Public Trustee, assigned the copyright to the record
company Larrikin Music.[39] Larrikin pursued legal action following the television broadcast, claiming that "Down Under" infringed its copyright for
"Kookaburra."

The Federal Court found against EMI Music, representing Men at Work,
deciding that the band infringed copyright by reproducing a substantial
part of the original work,[40] and the Full Federal Court dismissed EMI Music's
subsequent appeal.[41] By that stage, however, Judge Emmett questioned the
point that precedent and legal reasoning had led the court to:

> If, as I have concluded, the relevant versions of Down Under involve an infringe-
> ment of copyright, many years after the death of Ms Sinclair, and enforceable at the
> behest of an assignee, then some of the underlying concepts of modern copyright
> may require rethinking.[42]

The entire case becomes even more complicated when we take into account
what the courts could not: a rumor among many musicologists that Marion
Sinclair composed only the lyrics to "Kookaburra" and that the tune was
based on an "[old] Welsh folk song about a blackbird."[43] Therefore, much like
the sample-licensing system, we can also charge copyright—in terms of the
law itself and the jurisprudence around it—with being both reductionist and
ambitious. The objectification of copyright at a foundational level allows for
decisions to be made about these complex and perhaps unresolvable interactions between creative practitioners, cultural artifacts, and culture more
generally. As Judge Emmett notes, however, the decisions that result from
this jurisprudence do not actually solve anything but simply carve out a set
of poorly justified relationships between different actors while avoiding the
fundamental (and ongoing) complications and contradictions that sit at the
heart of copyright, particularly in reference to subjectivity.

Rebecca Tushnet makes an analogous point in a study of how US copyright jurisprudence treats images, arguing that the entire notion of a test for
substantial similarity in works becomes largely incoherent when applied
to non-textual works.[44] The ability to find infringement in works of substantial similarity requires courts to make substantive aesthetic judgments
about whether different artistic works look, feel, or sound different, and
she notes that these kinds of "gestalt evaluations" directly conflict with the
kind of "analytic dissection" required of courts and judgments.[45] Moreover,
conclusions drawn by fact finders to assist the court are likely to be heavily

influenced by advocacy, with the manner in which other people "talk about the work" directly affecting how the fact finders perceive the work itself.[46] Tushnet makes a convincing argument that these judgments, which fundamentally reshape how particular individuals are interpellated in relation to their creative work and the work of others, are based on poor reasoning that posits "an unsustainable dichotomy between unprotectable idea and protectable expression."[47] Although Tushnet's analysis focuses on images, the examples we have discussed show that comparable issues emerge when attempting to locate substantial similarity in music.

The preceding discussion also furthers the analysis of the way that copyright law and the music industry treat relational authorship. Authorship is granted to works that draw heavily on particular folk-based musical traditions such as folk songs ("Kookaburra Sits in the Old Gum Tree") or jazz and blues ("Got to Give It Up"). Once these initial moments of use are formalized through law, however, if subsequent works fail to get a license to sample music—that is, to draw on a musical tradition—or if they reproduce a "substantial part" of the existing copyrighted material (a vague and unclear standard), they are seen as infringing works. The problem with this structure is that it presumes the chain of influence between musical works can be clearly discerned and separated, which, as we have seen, is not always the case. For example, if we return briefly to the *Bridgeport* case ("Get Off Your Ass and Jam"), at what point does a two second guitar riff taken from George Clinton, transposed into a new key, and placed in a rap song become something entirely different? In a similar fashion, does the reproduction of the "Kookaburra" melody in a minor key, at a faster tempo, and in the context of a rock song by another musician stand as a transformative form of authorship? As Tushnet notes, these questions of aesthetic judgment are not that simple and perhaps should not be addressed as part of the rights that copyright purports to protect.[48] Whereas relational authorship is recognized through some authorized mechanisms (such as sampling), there is no deeper recognition of the complexities of authorship or the practical process of creativity, which entails numerous acts of use and borrowing between various authors and proto-authors.

The story of the flautist who played the infringing line in "Down Under," Greg Ham, also underlines the practical nature of these questions around subjectivity. When the Federal Court found against EMI Music, Ham felt like a thief and stated that the decision "destroyed so much of my song."[49]

Questioning his own legacy, he claimed that he would only be remembered "for copying something."[50] Here we see an individual repositioning their own subjectivity in relation to a court decision, moving from a space of creative production and authorship to that of an infringer, an unknowing user who was found to be a pirate. This poignantly demonstrates the extent to which the industrial and economic structures that allow relational authorship to operate can seriously affect how creative producers view themselves and their work.

Nonetheless, these limits around relational authorship are a genre-specific phenomenon and based in practice and culture as much as law. Indeed, as we now turn to the art world we see a different notion of relational authorship take hold, one that is conscious of industry practice to the extent that it challenges some of the foundational legal concepts that underpin copyright.

Contemporary Art and Relational Authorship

Artist's Assistants and the Morals of Creation

In January of 2016, I went to the Museum of Contemporary Art in Sydney to see a survey exhibition of Grayson Perry, the Turner Prize winning ceramicist. My visit aligned with a tour offered by the gallery, which gave some information on Perry's exhibit "Pretty Little Art Career." One factoid stuck with me; namely, Perry's ceramic work is so intricate and detailed that he is one of the few renowned contemporary artists solely responsible for the production of his own art. The most another person has done is help him move his large ceramic vases around the studio. This contrasts with the practices of many other contemporary artists, who lead teams of assistants that help construct and deliver the artwork. In some cases, the artist may not even produce any of the work but simply offer the idea and lend their name to the finished product.

For example, UK artists Gilbert and George hire assistants to color in massive prints that usually feature the two artists prominently. In an interview, two of their assistants explain that their jobs often involves coloring in "Gilbert and George's penises for eight hours a day."[51] Damien Hirst's art is produced in a similar fashion. Indeed, Hirst has admitted that he has "painted only five of the 1,400 spot paintings in existence" (see figure 3.1).[52] He is comfortable defending his authorship, however, explaining that:

Figure 3.1
Damien Hirst standing in front of one of his spot paintings. Credit: Andrew Russeth
(CC BY-SA 2.0).

[E]very single spot painting contains my eye, my hand and my heart. I imagine you
will want to say that if I don't actually paint them myself then how can my hand
be there? But I controlled every aspect of them coming into being and much more
than just designing them or even ordering them over the phone. And my hand is
evidence [*sic*] in the paintings everywhere.[53]

This explanation essentially refashions the artist, with their creative contribution looking like more that of an architect. Like architects, contemporary
artists are required to have the vision in the first place, direct the construction of the work, and sign off on the project. This is enough to embed their
authorship and "heart" in each work they produce.

Yet some architects still draw plans; that is, they engage in some labor
to create a material object that is copyrighted.[54] In contrast, although these

artists have a number of interesting creative ideas, their ideas are not copy-rightable and so their production practices do not square with this func-tional understanding of authorship. Therefore, production arrangements have to be organized through alternative legal mechanisms. In the United States, an assistant would transfer their copyright to the artist through the work-for-hire doctrine, which allows the transfer of copyright between par-ties when a work is created as part of an individual's employment.[55] Through this process, the employer becomes both the owner and author of the work. By contrast, in the Commonwealth jurisdictions examined in this book, an employment relationship will transfer ownership of the copyright and all attendant rights to the employer, but not authorship.[56]

There is not much difference between these two legal systems at a practi-cal level, as either system can transfer copyright from the artist's assistant to the artist, and the originating artist can either waive their moral rights[57] or consent to alterations that would otherwise breach those rights.[58] These sys-tems start to diverge, however, once we consider the construction of author-ship at a philosophical and cultural level. Although contractual agreements in the UK, Canada, and Australia transfer the ownership of copyright to the artist, the fact that authorship continues to reside with the artist's assistant at least purports to sustain a notion of sole authorship (albeit this presum-ably does little for the salaried assistant who has no economic rights in their work). Conversely, the work-for-hire provisions in the US allow for compa-nies to be interpellated as authors and owners, standing in for a prospective multitude of creators.

This might seem like a minor distinction but these two approaches offer different ways of thinking about multiple modes of authorship and how authorial practices interact and intersect with use. In a sense, the Common-wealth countries maintain a legal fiction in which authorship continues to reside in a sole subject, regardless of the fact that the only spot paintings of value to the art market are those "authored" by Damien Hirst. In con-trast, by allowing a company to claim authorship, work-for-hire provisions naturally support this kind of relational authorship under which cultural and industrial (and in the US, legal) authorship is granted to someone like Hirst who, in practice, simply "uses" and "names" the already authored spot paintings. Indeed, work-for-hire sustains one of the only feasible mod-els of authorship in the contemporary art world: the author as a brand.[59]

In the contemporary art industry where repurposed goods (consider Duchamp's toilet), mass-produced goods, and conceptual works in process[60] can be designated as works of art, the fiction of sole authorship cannot be sustained. Indeed, Amy Adler has noted that contemporary art directly challenges the successful functioning of moral rights regimes, which as noted in the introduction grant a series of personality rights to the author of a copyrighted work (and apply solely to visual works in the US).[61] Most continental and common law nations give the following non-transferrable rights to authors as part of their suite of copyright laws:

> The right of integrity, under which the artist can prevent alterations in his work; the right of attribution or paternity, under which the artist can insist that his work be distributed or displayed only if his name is connected with it; the right of disclosure, under which the artist can refuse to expose his work to the public before he feels it satisfactory; and the right of retraction or withdrawal, under which the artist can withdraw his work even after it has left his hands.[62]

Adler argues that the moral rights regime is completely at odds with authorship practices generally, but even more so in the US where it solely applies to visual arts, a sector that has increasingly embraced "depersonalized and commercial" methods of creation that rarely require the touch of the author's hand.[63]

At first glance, Adler is correct to point out that the moral rights regime interpellates an author subject that does not accurately represent how various industries approach authorship today. It is possible to interpret these rights from an alternative perspective, however, and use them to sustain the sort of relational authorship that work-for-hire provisions formally afford. Partly in response to Adler, Xi Yin Tang suggests that the notion of the author as a brand could be sustained by considering authorship in contemporary art through the lens of trademark law and using moral rights to protect brand reputation rather than authorial reputation. Tang agrees with Adler that the artistic production lines set up by the likes of Hirst mean that originality can no longer necessarily be defined by "novelty, artistic merit, skill, or bold, artistic gestures."[64] Instead, the artistic brand has become synonymous with "the deliberate control an artist exerts over his body of work in order to create an identifiable brand for himself [replacing] personal connection as the hallmark of contemporary art-making."[65] Although Hirst might claim a personal connection with each spot painting, it is not clear that his overarching claim of authorship should override that of the assistants who actually painted the work. What he does hold a clear claim over, though, is the Hirst brand.

Tang explains that artists could make use of moral rights to protect their work. She suggests that these rights (in particular, the right of integrity) could be deployed in order to maintain brand authenticity. Under various moral rights regimes, reputation has often been interpreted in relation to the sole person of the author, but Tang argues there is no reason that this should be the case.[66] In the same way that trademark law allows us to confidently identify a Louis Vuitton item or a Mercedes, protecting the brand from trademark dilution (i.e., competitors attempting to use a similar trademark that devalues the uniqueness of the original) or passing off (i.e., the protection of a brand's reputation against attempts to damage it via association), moral rights could be used to protect the "brand" of an artist. Furthermore, when an artist protects their reputation, they are not just protecting themselves but also "other owners of [their] work by ensuring that the brand they own stays unique and untarnished."[67] This argument shows how a broad reading of moral rights could function as an important method of protection for authorship and also gives us some insight in how to think beyond existing models of authorship in copyright.

Tang's proposal is just one solution to the authorship problem. It might help consumers, intermediaries in the art market, and the likes of Damien Hirst, but it may not offer much succor to the artists' assistants who have had the prospective financial benefits of their creation transferred either through the work-for-hire doctrine or via contract in exchange for a salary. Of course, some assistants are happy to claim that their work is not creative or authorial, as one of Gilbert and George's assistants argues:

Does the person who makes the hubcaps or whatever they're called these days—low-profile sports rims—point at a passing Mercedes SLK or whatever it's called, saying, "I did that?" No. So why should assistants claim possession for their work? It's a job.[68]

Conversely, an assistant for sculptor Tara Donovan says that she does "think about not getting credit," particularly as "it is very hard after eight hours of constant work in her studio to come home and push yourself to do your own work."[69]

To briefly turn to another creative sector, similar issues around the accurate and fair identification of authorship have emerged in the film and television industries. With a myriad of parties involved in the authorship of screen content, from actors and directors to writers and producers, collectively bargained residual payments organized through directors or writers guilds are regularly used to share income generated from the production.[70] Coproduction credits are also strategically used to grant control of copyright ownership to particular parties deemed critical to the creation of the work.[71] What does it say about copyright law that contracts, the work-for-hire doctrine, guild agreements, and strategic credits are required to evade some of the foundational logics of copyright?

At a minimum, these industry practices emphasize that copyright can often complicate rather than clarify authorship due to the limited nature and function of the rights in a collaborative environment (particularly in Commonwealth countries where authorship is not immediately transferred to an employer). Moreover, the examples discussed in this section also highlight the need to think about authorship in an entirely different fashion, particularly with reference to moral rights. With authorship claimed by artists who essentially appropriate entire works created by their assistants, the lines between the roles of author, user, and pirate at a conceptual (rather than contractual or legal) level become increasingly blurred in Commonwealth countries. As noted earlier, the work-for-hire provision at least offers doctrinal recognition of the sort of collaborative environments in which copyrighted works are often produced, and the mechanism also works well with the reconsideration of author as brand posited by Tang.[72] Yet, as discussed, both Commonwealth and US jurisdictions can fail to accurately distribute finances and discern authorial contributions in their own ways. Subsequently, the role of intermediaries in establishing (and denying) authorship also forms a critical part of how relational authorship can be sustained and maintained.

Free Expression and the Fair Use Conundrum

Relationality appears in a more provocative fashion around questions of use, reproduction, and originality in art. In this respect, art again functions in a completely different fashion than the music industry, which, as we have seen, requires a payment for the use of copyrighted material, no matter how small. In contrast, some pieces of appropriation art have been able to draw heavily on existing material and still be considered transformative by the courts. It is worth considering the practice of Jeff Koons, an appropriation artist in the US who has repeatedly been sued for copyright infringement. Originally, in the early 1990s, his artistic practice did not find much favor within the courtroom and many of his works were found to be infringing copyright. One involved using an existing photograph of "a couple holding a litter of puppies"[73] and another involved the production of a sculpture of *Garfield* character Odie as part of a larger piece.[74] In both cases, his artistic practice was not considered transformative enough to warrant a fair use defense.

A later appropriation art series was more successful in establishing itself as transformative. Koons produced collages of advertisements and his own photographs and then superimposed these images "against backgrounds of pastoral landscapes"; the images were then used as templates for his assistants to refer to when producing the finished paintings.[75] These works formed the series "Easyfun-Ethereal," a collection that referenced particular ads as well as popular culture more generally.[76] One of the paintings featured the work of fashion photographer Andrea Blanch, who had taken a photo of "a woman's lower legs and feet, adorned with bronze nail polish and glittery Gucci sandals, resting on a man's lap in what appears to be a first-class airplane cabin."[77] Koons had used a section of the photo encompassing the lower legs and feet for the work "Niagara," which formed part of the series.[78] Blanch went to the Guggenheim Museum in 2002, happened to see the offending painting, and filed suit in 2003, accusing Koons of copyright infringement. The US District Court for the Southern District of New York found that Koons did not infringe copyright due to the fact that his use was transformative and thus allowed under the fair use exception.[79] Blanch appealed but the Court of Appeals for the Second Circuit also found for Koons.[80] The court decided that "Koons had a genuine creative rationale for borrowing Blanch's image" and that his "own undisputed description" explained that the image was being used "as fodder for his commentary on the social and aesthetic consequences of mass media."[81]

What is interesting about this case is how copyright law's consideration of visual art differs from its consideration of musical works. As we have already seen, both copyright law and music industry practice have entrenched a permission culture in which individuals are required to get a license or stop sampling music, no matter how small the piece may be (unless the work is a parody or falls under another copyright exception).[82] Potential creators are viewed as users and pirates who need to make a request to transition into an authorized authorial subject. In contrast, Koons has been granted greater leniency to reproduce a significant portion of a copyrighted work, even though he did not seek a license. Moreover, he was given the position of a user–author by the court and recognized as an individual who can "progress science and the useful arts" even though he relied on assistants to produce a significant proportion of his copyrighted work. This decision must be read in line with an increased recognition of postmodern artistic practices that drew extensively on previous works.[83] Although the law did not recognize Koons's earlier practices as artistic, over time the courts developed a cultural awareness of postmodern art and the forms of pastiche and bricolage that supported the genre.

Copyright jurisprudence in the US promises to consider each fair use case on the facts and not simply enforce "bright-line rules."[84] Still, there are nuanced cultural tendencies that get subtly promoted in these judgments. Decisions are based on precedent but they also speak to long-held ideas of visual artists as people who must have the right to maintain their freedom of expression, to be able to comment on current trends, and to function as lightning rods of progress. Rightly or wrongly, commercial musicians—at least if we consider the jurisprudence outlined in this chapter—do not seem to be held to a similar standard and function with stricter rules around how and when they can use existing copyrighted work. Of course, Koons has not won every legal battle he has fought. The point, though, is that the transformative nature of a new use has been given much greater weight in cases considering visual art.

These differing copyright decisions also have something to do with the specific nature of the art object and the structuring values, beliefs, and practices of the art world. Recorded music has historically operated as a stable commodity with a fixed value that has been easy to control from centralized locations such as record company headquarters. In contrast, the value of an art object can fluctuate rapidly as "prices, price differences, and price changes are associated with changes and differences in [opinions about the] quality of the work," lending a highly subjective and aesthetically oriented element to pricing decisions.[85] The art world also thrives on challenging artistic practices, and novel approaches (like that of Koons's) are welcomed and promoted by leading figures.[86] These embedded tendencies mean that art objects cannot be regulated by an industrially organized transactional licensing system in the same way that recorded music is. Moreover, even if such a system were possible, the unique demands of the art world would lead auction houses, collectors, and galleries to take risks favoring new, innovative, and potentially infringing artworks.

Yet we must also keep in mind Adler's recent study of fair use in contemporary art, in which she notes that the transformative use test has failed to support artistic practices based on copying.[87] Furthermore, jurisprudence in this area has not provided clear guidelines on what actually counts as transformative with regard to visual arts.[88] As well as a useful caution to the narratives of fair use triumphalism surveyed in the previous chapter, this study also signals that judicial tendencies can still have a severe effect on artistic practice despite the noted bias toward the artistic need for free expression. That bias still holds, Adler notes, as the artist's statement (in which the artist explains the intent behind their work) can play a critical role in a finding of transformative use. Indeed, she contends that *Blanch v. Koons* was only successful because "Koons had learned how to testify in a way that pleased the court," referencing the "new insights" his work would offer the viewer.[89] Adler says that she, too, would have "advised any artist worried about using copied material … to record clear, contemporaneous statements of any transformative intent."[90]

Courts, therefore, must clearly still welcome the artistic right of free expression, but it must be framed appropriately. This points to another cultural issue that emerges around the boundaries of authorship, use, and infringement, namely, that the courts expect free expression to go hand in hand with artistic intent despite the fact that the latter is a useless category in an era in which there is often "no artist, or … multiple artists," leaving fair use to rely "on a criterion that contemporary art long ago abandoned."[91] There is still a comparative permissiveness toward the use of copyrighted artistic works as opposed to musical works, but Adler's study shows that copyright continues to struggle with the kind of relational interactions that creativity requires. Indeed, the limitations around the use of musical quotations, the complexities regarding authorship of artistic works, and the need to use contracts to avoid fundamental issues in copyright doctrine demonstrate that copyright law is unable to produce a coherent relational author across multiple industries.

Industry Practices and Policy Solutions

This chapter has provided further evidence to support one of the central arguments of the book: that copyright law recognizes relational authorship at a basic level. The extent of this relationality, however, is defined by cultural expectations of the roles that particular artists should play within society, the expectations and norms of each creative industry, and copyright doctrine. The music industry recognizes relationality only through the lens of economic transactions, and courts have relied on an arguably illogical test to make distinctions between authorial and infringing acts in musical creation. Visual artists are given more leeway to make use of existing copyrighted works but jurisprudence around fair use illustrates that the courts have particular cultural expectations about what art can and should do when in engaging in this process. This chapter's examination of authorship in contemporary art has also shown that although the work-for-hire provisions in US law create a productive space in which relational authorship can occur, legal doctrine is often unable to accurately discern authorial contributions and distribute finances, which results in the use of intermediaries or formal contracts to manage authorship.

I have also welcomed the idea of thinking about certain authors as brands, and have noted that this concept is already operational (to an extent) through the work-for-hire provisions. This doctrine not only provides a space for relational authorship but also avoids some of the awkward conceptual (as opposed to practical) contradictions that have emerged in Commonwealth countries when ownership is transferred to an employer. In truth, it is not a particularly radical idea. The history of authorship is essentially the story of the author being used as a market device that allows particular creative practices to coalesce around the subject in an attempt to ease the flow of commerce in particular creative industries. One can occasionally discern a clear relationship between the process of creation and authorship, but at other times the author stands in for beneficiaries, from the Stationers' Company to the contemporary art market. Subsequently, this notion of an authorial brand is not much of a historical anomaly. The author has always been used as an avatar.

Indeed, what is perhaps more of a problem for copyright is the presence of contracts and union agreements, which directly challenges the scope and effectiveness of copyright as a policy instrument.[92] Copyright law promises a sustainable and functional philosophy of creation, creative progress, and due recompense, but instead a number of creative industries are avoiding copyright in order to better manage the constellation of personal debts, quiet agreements, small resentments, and formal contracts that surround the concept of the author. This suggests that claims regarding the effectiveness of copyright law in sustaining creative industries need to be tempered, as more often than not copyright is only one piece of a more complex puzzle around how industries determine authorship and generate an income. In closing, I propose that it might be best not to criticize copyright for failing to offer the perfect conception of a relational author or failing to address the needs and wants of complex entertainment industries; instead, we could advocate for copyright practitioners to recognize the limitations of the author subject in copyright, particularly in relation to the development of copyright policy. If the sort of relational authorship that supports many creative industries cannot be embedded in doctrine, at least the "policy narrative" of copyright could avoid simplifying the complex industrial, cultural, and legal processes that help establish and define the boundaries between authorship, use, and infringement.

4   Locating the User and Reforming Law

Who is a user? Tech industry professionals use the word to describe people who do things with technology. Intellectual property lawyers use it to describe actors engaging in authorized activities with copyrighted works. These naming conventions have unsurprisingly influenced how scholars use this term, and the word is common among researchers working in related areas for the same descriptive reasons. Even though "user" has entered the cultural zeitgeist, however, the subject is still an under-examined figure. We still call collectives of people "users" in an offhand manner when we are describing incredibly diverse people who may be engaging with the same media (or other copyrighted material) in a range of different ways for a multitude of reasons.[1] The discipline of science and technology studies (STS) has long recognized this problem and come to a deeper understanding of how different people use technology (as we will see later on). The same cannot be said for the discourse (whether scholarly or otherwise) around copyright law and digital media.

This lack of recognition is a strange phenomenon because the user has never been more noticeable. As we saw in chapter 2, the user has become increasingly visible thanks to the growth of amateur creativity and the mainstream adoption of social media. Marketers laud user-generated content (UGC) and scholars analyze the phenomenon of amateur creation. Everyone is attempting to better understand these new ways of producing and consuming.[2] Even copyright law has progressed from its longstanding focus on the author. User rights are no longer just the concern of digital rights groups like the Electronic Frontier Foundation but are now an issue taken seriously by a number of common law legislatures. Indeed, as we will see, the subject has been granted a range of new legal rights following reforms in Canada, Australia, and the United Kingdom.

We already know that users benefit from exceptions and limitations to the exclusive rights granted by copyright, and these new reforms tend to be presented as simple extensions of existing user exceptions. In some cases, however, we see an attempt to radically reconsider the scope of a user's ability to use copyrighted works. At first glance, this might appear to be a good thing. For a number of years, digital advocacy groups, librarians, and copyright scholars have been demanding greater freedom for users and a fairer, less author-centric approach to copyright law, one that reflects the realities of today's changed media environment. But things are not as simple as they seem. This general advocacy of user rights has avoided some of the more complicated questions around subjectivity and copyright: Who do we invoke when we talk about users? Who gets to define this subject position? And, how useful is this word as a descriptive term anyway?

In short, many discussions around copyright law and digital media talk about users as though they are legible subjects when they are anything but. Julie Cohen notes that the category is regularly described in a "highly heterogeneous" fashion in various works of copyright scholarship and argues that even when scholars promote different ways of conceptualizing the user, none of them are truly satisfactory.[3] Cohen offers the productive concept of a "situated user" as a way forward (briefly discussed in the introduction), which I will examine in more detail later in this chapter.[4] This is really only an early foray into what will be some heavy conceptual work for copyright scholars (a process that has already begun with Kylie Pappalardo's detailed conceptualization of the user).[5] With copyright, defining a user becomes even more difficult because in practice the category also captures libraries, galleries, and other public institutions as well as intermediaries like record companies (for example, when licensing samples).[6] Not that disciplines such as Internet studies or media and communications are any different. In those fields, the user is regularly "presented as a neutral actor in the contemporary digital moment" and so is often constituted through a particular strain of technological determinism.[7]

This chapter takes a step toward unpacking the user by exploring the different ways the subject is positioned in copyright's relational triad in reference to the author and the pirate. I begin by outlining the recent extension of user rights in Australia, the UK, and Canada. I then examine the Canadian reforms in more detail, as they have taken the most radical approach to user rights. In addition to retaining a complex relationship with foundational doctrinal concepts of creativity and authorship, the reforms also raise questions around how to talk about this phenomenon of user creativity and the best way to enforce copyright in light of its preponderance. I then look at how intermediaries contribute significantly to the construction of the user as well as the facilitation of use. I analyze two cases in which Australian and US courts impeded commercial content distributors who had strategically organized their operations around particular legal protections afforded to users, suggesting that particular commercial activities performed by intermediaries cannot be incorporated in copyright's construction of the user. I then return to Canada and explore how particular intermediaries are afforded the flexibility of user exceptions as they are seen to directly assist users and go on to highlight the importance of consumption during a cultural moment that celebrates the creativity of users. The chapter ends with a reflection on the best way to think about the user in a relational light, drawing on research from law and STS to sketch out new analytic trajectories.

New Directions in Legal Reform

The extension of user rights over the last decade or so has had much to do with the fact that many jurisdictions have a specific set of user exceptions embedded in law (discussed previously in chapter 2). As noted there, without the benefit of a broad fair use provision (as in the United States), a use becomes an infringement of copyright if legislators have not anticipated it. Because law is unable to easily keep up with recent technological developments, this creates a major problem: if a particular practice catches on (such as the now historical practice of transferring music from a CD to an iPod), the bulk of the population can (unknowingly and without malice) infringe copyright on a regular basis. Numerous countries have reformed their exceptions to copyright in order to reorient it to the increasingly mainstream uses of digital media that were clashing with a law structured around industrial (rather than networked) modes of production and consumption.

In Australia, a suite of user-oriented reforms was introduced in 2006.[8] The changes were introduced in part because a recent free trade agreement with the United States seemed to "tilt the balance towards copyright owners to an unacceptable degree."[9] As suggested previously, however, the key driver behind these reforms was the broader cultural shift occurring with regard to technology use (most notably the mainstream adoption of the iPod), which had changed the scale and nature of private copying. Prior to the reforms, Australian copyright law already had user exceptions for educational purposes, criticism, news reporting, or professional advice under fair dealing.[10] But it did not protect the most basic forms of content manipulation, private copying, and content storage, let alone the extensive forms of remix being showcased on the Internet.[11] This led to a paradoxical situation where iPods were being sold in droves to enthusiastic consumers, despite it being illegal to transfer any copyrighted music from a CD to a computer and then to the machine itself.

In addition to addressing the issue of format shifting, the Australian government used this opportunity to tackle piracy at the same time, which led the attorney general to frame the amendment as making things "fairer for users and tougher for pirates."[12] Things were made "fairer" for users in a number of ways. The Copyright Amendment Act 2006 legalized certain kinds of format shifting, meaning Australians could reproduce music and books in different formats for private use, albeit in very particular circumstances. Time shifting, which was already arguably legal because of how Australia's authorization law had been interpreted, was also formally codified and its legality in relation to digital contexts was confirmed.[13] Finally, parody and satire were added to the existing list of fair dealing exceptions, legalizing the culture of humorous remixes that were flourishing online.[14] In addition to this generous reform package for users, however, a number of strict liability offenses having to do with copyright infringement were introduced, as well as a wider range of enforcement options for relevant authorities.[15]

The United Kingdom's reform process was sparked by a similar set of issues. Ian Hargreaves conducted a review of the intellectual property system in 2011 and found that "exceptions have failed to keep up with technological and social change, leading to widespread consequences."[16] He recommended (among other things) that the UK adopt a private copying and format-shifting exception that "corresponds to what consumers are already doing."[17] This would allow individuals to legally use CD burners and MP3 players and "make copies for their own and immediate family's use on different media."[18] A parody exception was also proposed, with the review noting that "[v]ideo parody is today becoming part and parcel of the interactions of private citizens," since UK copyright law still allowed rights holders to remove what were fair and humorous parodies (albeit with limited success).[19]

The UK government went on to conduct a consultation process and a multitude of views around private copying emerged.[20] Consumers and tech companies generally supported the recommendation but representatives for copyright holders did not. In a way, this opposition was academic. Private copying was widespread and the existing law was largely unenforced. The parody exception also drew contrasting arguments. Some saw it as a useful way to legalize user-generated content that was already appearing online, but other respondents argued that the exception would make comedians effectively "immune from paying royalties to incentivise the creation of works."[21] Despite these contrasting views, after the consultation the government went on to broadly accept the recommendations made and the private copying and parody exceptions (among others) were approved by Parliament in July 2014.[22] Still, these exceptions were not passed without some debate. There was consternation that the government did not pay some sort of compensation to copyright holders or enact an accompanying private copying levy. Some in the House of Lords suggested that the government was "putting the creative industries at risk,"[23] and industry group UK Music criticized the lack of "fair compensation."[24] Although the exceptions were enacted, the lack of a levy eventually proved to be fatal for the proposed private copying exceptions. UK Music (along with the Musicians Union and the British Academy of Songwriters, Composers and Authors) won a subsequent court challenge in 2015 that quashed the exceptions, meaning that private copying, "which would have fallen under the exception, now constitute[s] acts of infringement."[25]

Canada has undertaken perhaps the most novel attempt at extending the rights of users. The Copyright Modernization Act, enacted in 2012, features a number of user exceptions similar to the reforms previously detailed.[26] The act legalized time shifting and format shifting, placed "a cap on statutory damages for non-commercial infringement" and also enacted "a broad series of exceptions facilitating greater ease of use for educational purposes."[27] It also introduced an exception for UGC, allowing an individual to use "an existing work or other subject-matter or copy of one ... in the creation of a new work or other subject-matter" as long as certain restrictions are followed.[28] This type of use is allowed providing: it is "solely for non-commercial purposes"; the original source is mentioned, "if it is reasonable in the circumstances to do so"; the individual has grounds to believe that the existing work is not infringing copyright; and

that the new work "does not have a substantial adverse effect on the exploitation or potential exploitation of the existing work ... or on an existing or
potential market for it."[29]

The UGC exception is broadly constituted, which gives Canadian users
a significant level of protection. Although the parody and satire exceptions
enacted into UK and Australian law sought to address the issue of amateur
creative work that drew on existing copyrighted material, those exceptions
have been positioned in relation to humor. They will not protect users who
may draw on existing copyrighted work in a non-humorous but interesting
fashion. In contrast, the Canadian UGC provision established "a legal safe
harbour for creators of non-commercial user generated content," which,
alongside parody, can also include "remixed music, mash up videos" and
even "home movies with commercial music in the background."[30] Therefore,
this exception stands as perhaps the best attempt by a jurisdiction to come to
terms with online amateur creativity. Rather than introduce a narrow exception addressing merely one element of online creative practice, the UGC
provision attempts to intervene in, and subsequently regulate, various forms
of user creativity by setting out general baseline legal standards for acceptable uses of existing copyrighted works.

Apart from this last provision, most of the Canadian reforms are similar
to those of the UK and Australia, and taken together these three moments of
reform can tell us something about how the user is viewed. Broadly speaking, these three jurisdictions were looking to legislate for users that wanted
to manipulate content: these were technologically literate people who time
shifted and format shifted their content and occasionally used copyrighted
material for creative acts. The legislation involved pushing back against
claims from copyright holders and industry groups that time shifting or
format shifting could damage future markets. Furthermore, these reforms
recognized that there was a relationship between use and creativity, but it
was managed in slightly different ways. On one hand, the UK and Australia
offered an exception for parody or satire, which suggested that these countries viewed the user as somewhat creative but still more of a consumer
than a full cultural contributor. On the other hand, the Canadian reforms
appeared to recognize the creative agency of users, that is, the relationship
between use and creativity, by offering an exception that extended the relational phenomenon of an authorial user.

At this point, it is worth noting that these exceptions are all limited by the three-step test found in Article 9(2) of the Berne Convention for the Protection of Literary and Artistic Works. I do not want to spend too much time examining international law but instead simply note in closing this section that a broader network of treaties and conventions ultimately sits behind these national jurisdictions and places real (albeit somewhat flexible) boundaries around the scope and capacity of national laws to make radical changes to the legal construction of the user. The article sets out the broad conditions on which a signatory can offer exceptions or limitations to the right of reproduction. The exceptions can occur in "certain special cases, provided that such reproduction does not conflict with a normal exploitation of the work and does not unreasonably prejudice the legitimate interests of the author."[31] Following the outcome of a World Trade Organization dispute case in 2000, some scholars have interpreted the three-step test relatively strictly and argued that it "does not give judges sufficient latitude for considering other interests than the right-holders."[32] In contrast, others have suggested that the test is being interpreted in an increasingly flexible manner and, indeed, Peter Yu suggests that the Canadian UGC exception would pass the three-step test.[33] Still, the fact that the test seeks to create a framework from which exceptions and limitations can be enacted fairly is worth thinking about because, as the following analysis of the UGC exception will show, it is not always possible to embed a relational user in copyright law equitably.

What's in a Name? Regulating User-Generated Content

At first glance, the Canadian UGC exception appears to be a clever piece of legal reform. Copyright law needs to respond to the challenges posed by the Internet and digital media more generally, so it is hard not to instinctively welcome a reform that directly addresses the phenomenon of amateur or user creativity by taking a broad and generally flexible approach. If we look at this amendment more closely, however, it becomes clear that despite its attempt at embracing a relational approach to subjectivity, it entrenches a hierarchy around what kinds of creativity can be commercially exploited.

The provision places a number of restrictions around the use of copyrighted works for UGC. Most notably, the exception allows people to engage in this sort of use for non-commercial purposes only. As Teresa Scassa notes, the exception "perpetuates the myth that the regular 'creator' does not borrow from or use the works of others [and] constructs UGC as a more parasitic activity than perhaps it deserves to be."[34] I have also noted elsewhere that "the exception immediately discounts the transformative elements of actions like remix, mash-up or the deployment of professional content in amateur settings."[35] Although this exception cannot stop someone from building on an existing copyrighted work to create a new original work, it discourages a user–author from exploiting this new work commercially, inadvertently setting boundaries around authorial creation. Under this exception, "real" authors can commercially exploit creations whereas "users" can only "generate content" in a non-commercial space.

An additional exception reinforces this creative hierarchy. Paragraph 29.21(1)(d) allows users to create works drawing on existing copyrighted material only as long as the new work does not have a "substantial adverse effect financial, or otherwise, on the exploitation or potential exploitation of the existing work."[36] This sort of creation is therefore still "subject to the whims of the rights holder through this 'open-ended' exception."[37] Scassa explains that rights holders could use this exception to argue that a particular use "diminishes the cultural impact or significance of the [original] work by trivializing it, or ... tarnishes the reputation of the [original] work as, for example, where fan fiction strays into the pornographic."[38] Once again, this exception places the creativity of "users" under significantly more scrutiny than that of "authors." Whereas the authorial creativity defended by copyright law requires only the production of a work that is original and fixed in a material form, under these UGC provisions any creativity that rests on the use of copyrighted material has to negotiate a number of exceptions and runs the risk of drawing the ire of the original copyright holder.

The UGC provision links the user with the author by recognizing that users can create and by developing a pathway that legalizes this sort of "secondary" creation. Yet this sort of creativity is not seen as equivalent to "authorial" creativity. The way the exceptions are structured implies that authors are engaged in an autonomous creative process that develops new original works, whereas users essentially "take" from existing creative works in a fashion that could directly threaten the work's market or reputation. Therefore, this sort of positioning also links the user to the pirate because, lacking a broad protection for use, users are defined by the threat of copyright infringement: They can take some copyrighted material, but not *too* much, and it has to be used in the right context.

These issues around the provision are not due to specific failings embedded in the Canadian reform process but the fact that we still do not know much about the user as a subject. This critique of the Canadian UGC provision is really just a way of opening up a discussion around the difficulties of understanding and legislating for this subject. Vague neologisms like UGC do not help matters. The phrase presumes we can already locate a stable user subject who is engaging in a series of clearly identifiable practices of content generation. When it comes to actual policy discussions around users, however, these terms and phrases show themselves to be highly contingent and defined in relation to the particular political stance of each stakeholder.[39]

Kristofer Erickson has made this point with reference to the European Commission's (EC) 2014 consultation on copyright, which was the commission's own attempt at addressing the phenomena of user creativity.[40] Throughout those policy discussions, there was not one stable definition of a "user." Authors' groups viewed users as a threat to professional creativity and positioned UGC as an entirely different practice than artistic creation, in a fashion similar to the Canadian exceptions discussed earlier. Service providers and platform operators (such as *Wikipedia*, Google, and cable and mobile phone operators) broadly welcomed UGC because of its "economic value," but also because of "its contribution to democratic principles and free expression."[41] But most noticeably, and in addition to advocating for broad rights to engage in transformative use, user rights groups directly challenged the EC's definition of UGC. They noted that UGC produced a false hierarchy of creativity, ignoring the fact that "transformative works constitute original work even as they borrow material from previous works."[42] Furthermore, this language forcibly separated "the amateur 'user' from the 'professional' author" when that distinction did not even "exist in copyright law."[43]

These divergent positions show that rather than being a stable activity that can be identified and legislated for, UGC is in fact "politically contingent and ideologically constructed in the context of the copyright reform debate."[44] This is partly because users themselves are represented in various ways, which in turn means that there is no way to establish a "stable and objective definition of online creativity."[45] We have already seen this issue emerge during the UK government's consultation process and, indeed, a similar issue occurred in Australia with stakeholders presenting three competing and largely incompatible visions of the user during the reform process.[46] Erickson suggests that this sort of definitional flexibility can be "politically advantageous," as it allows for the issue of UGC to function as a central point of debate around copyright law more generally.[47] Rights holders can "highlight the potential for direct infringement and piracy" and "platform owners and users can evoke the creative and yet non-commercial status of UGC," with UGC standing in as a productive proxy for a number of hot-button issues.[48]

Still, it is difficult to see how useful this sort of definitional uncertainty is for people who need to rely on exceptions structured around a particular notion of UGC. This uncertainty also underlines the fact that both the phenomenon of online creativity and the user subject itself are not obvious or legible but rather fundamentally political constructs. The fact that debates have occurred in multiple jurisdictions around who is a user, how their creative outputs should be regulated and understood in relation to authorial creation, and the extent to which this creativity can be exploited economically proves this point. These recent user reforms have raised a lot of interesting and difficult questions but have not provided any direct answers about how the user should be constituted in relation to developments across digital media more generally.

The undetermined nature of UGC also means that any reforms that are passed tend to be designed for a simple imaginary of a user while ignoring some of the more complex ways the user is constituted. The parody and satire reforms in the UK and Australia and the general UGC exception in Canada ultimately legitimized creativity that drew on copyrighted works but did not directly affect the existing markets of the works. Similarly, although the format and time shifting exceptions were necessary reforms, they still approached the user as essentially a consumer of copyrighted works. Amateur creativity is not necessarily a non-commercial practice, however, and with social media economies increasing in sophistication over the past few years, practices that have been labeled "UGC" are looking less like idealized non-commercial forms of user creativity and more and more like commercial authorial acts.

YouTube, MCNs, and the Let's Play Phenomenon

YouTube is no longer an online platform that distributes amateur video productions (see figure 4.1). It is an increasingly professionalized environment.[49] Entrepreneurial amateurs are regularly being signed up by multi-channel networks (MCNs) who target popular YouTube channels across "every conceivable niche" from gaming and dancing to instructional beauty videos.[50] MCNs often represent tens of thousands of clients, which means that one of their most important jobs is to build back-end tools for creators that can help them "manage IP, create thumbnails, and source stock footage."[51] They can also work directly with YouTube creators to "develop branded content, typically in the form of product placement, integration, and themed videos around consumer brands."[52] In exchange for this support MCNs receive a cut of YouTube creators' ad revenue. Although people might be producing content that "looks" amateurish (and even

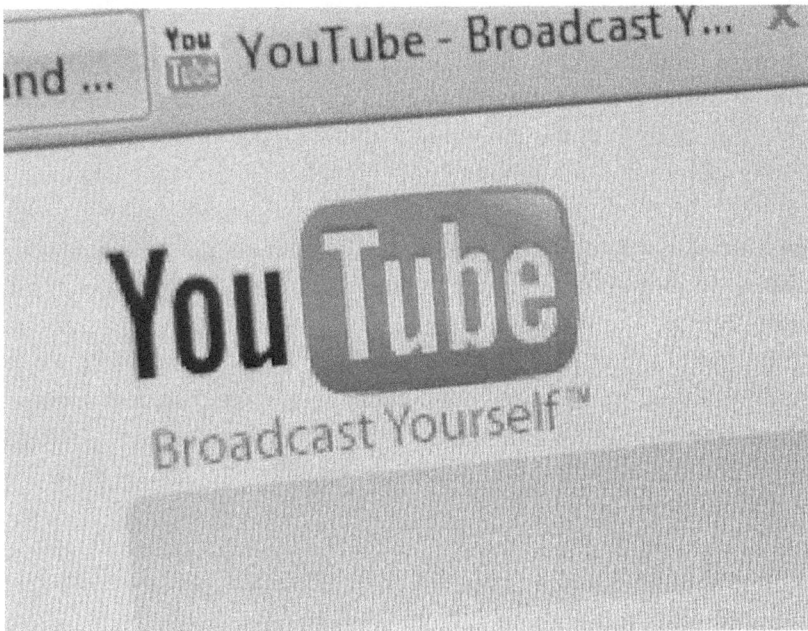

Figure 4.1
YouTube has become an increasingly commercialized platform and has encouraged the monetization of formerly amateur content. Credit: Andrew Perry (CC BY-SA 2.0).

this aesthetic is changing), the tensions around UGC become more readily apparent when we consider the fact that this sort of production is occurring in an increasingly commercial environment.

Ongoing commercialization has had real consequences for popular forms of amateur creativity, which draw heavily on copyrighted works. For example, watching video game playthroughs, known as "Let's Play" videos, is one of the most popular uses of YouTube today. "Let's Play" videos feature someone playing a video game to completion and commenting as they are doing so by talking to the camera. The leading practitioner of this genre, Felix Arvid Ulf Kjellberg (a.k.a. PewDiePie), was an established YouTube star until recently.[53] These practitioners need to use an extensive amount of existing copyrighted material in order to produce this content and whereas some video game companies are happy to enjoy the free publicity that stars such as PewDiePie deliver, others are less happy to see people earning money from YouTube ads off the back of their copyrighted content.

The most notable response to this phenomenon has come from Nintendo, which registered as a YouTube partner in 2013. This allowed the company to place ads in videos that featured their content.[54] Nintendo then ramped up their approach in 2015 by releasing the Nintendo Creators Program.[55] YouTube creators were able to sign up with Nintendo and receive official authorization for their use of "Nintendo-related content," but they would then have to give up 30 percent of any advertising revenue from across a registered YouTube channel and 40 percent of any revenue made from individual videos.[56] Although this could be seen as something of an olive branch to "Let's Play" creators, many have responded negatively to the program, arguing that their videos offer free publicity for Nintendo products and noting that Nintendo would take ad revenue from their entire channel, even if some videos did not feature Nintendo content. Nintendo also has to approve videos before they are released and "can alter the terms of the deal" at its leisure.[57]

Furthermore, by registering as a YouTube partner Nintendo was also able to use YouTube's infamous Content ID program. Content ID compares uploaded video footage against a database of files submitted by partners. If a match is found, partners are able to mute a video's audio track, block the video entirely, place ads in the video to monetize it, or simply track the video's statistics. In addition, these actions can be country-specific, so partners could block a video in one country while monetizing it in another.[58]

Of course, one of the major issues with this system is that it is largely automated. Once Content ID makes a "match," a partner like Nintendo has the power to enact any of the above choices and it is up to the video uploader to appeal the decision.[59] In a similar fashion to previous technologies such as digital rights management (DRM) and technological protection measures (TPMs), Content ID purports to offer an easy solution to copyright infringement but "the translation of legal rules into code" has not proven to be "particularly adept at handling copyright's legal subtleties."[60]

Surprisingly, YouTube has recently performed something of an about-face when it comes to enforcing copyright. The company (and its parent Google) has been a strong supporter of Content ID since its launch, arguing that the tool's flexibility allows copyright holders to make the best decision on how to protect their copyright.[61] It has maintained this line even in response to criticism that giving the identified copyright holder the power to take down content (as opposed to the individual uploading the video) has led to the removal of numerous videos that should have been seen as "fair use."[62] This position changed dramatically in 2015 when Google announced that it would "protect ... the best examples of fair use on YouTube by agreeing to defend them in court if necessary."[63] Google naturally framed this endeavor as a positive intervention into the copyright debate but industry analysts noted that the announcement came soon after the launch of YouTube's paid subscription service YouTube Red.[64] This launch changed YouTube's strategic orientation, as the subscription service would focus on original programming rather than operate as a repository for mass-market videos. Therefore, the platform's copyright strategy would have to change as well, from acquiescing to large copyright holders to protecting its established stars who rely heavily on the reuse of existing copyrighted content.[65] This turn of events highlights the financial and strategic importance of user–creators and the content they produce. They are generating so much income for YouTube that they are considered to be critical to the future financial viability of the platform.

In examining UGC, we have taken something of a circuitous route, starting with a critique of the Canadian UGC exception, traveling through the European Commission's attempt to come to grips with UGC, and ending with a discussion of the professionalization of YouTube and how copyright is managed on the platform. What we see across all these examples is that both user-generated content and the user subject are profoundly unstable political, cultural, and legal constructs. We saw that in both Canada and Europe, UGC has been positioned within a questionable hierarchy that views "authorial" creativity differently from creativity that draws on copyrighted works. This in turn carries implications for how the user is viewed. Both Canadian legislation and authors' groups in the EC have positioned the user as a somewhat parasitic or piratical figure and have sought to either drastically limit user-oriented reforms (in the EC) or limit to the non-commercial sphere any UGC that draws extensively on copyrighted works.

After looking at what was actually happening with UGC, however, we saw a user subject that did not look much different from a professional author. People are creating content in a commercial context, signing contracts with MCNs who function as highly networked and distributed agents, and even have the option to negotiate the use of copyrighted content with copyright holders (albeit in contentious circumstances). This suggests an alternate picture of the user as one who has the agency to create content and function in an authorial fashion. We are also (once again) seeing a situation where the very infrastructure of copyright law is being ignored. Ventures like the Nintendo Creators Program manage use through contract rather than in reference to copyright doctrine. Similarly, Content ID claims to support copyright but functions in a much different fashion that does not necessarily account for the nuances of use that (in an ideal world) are supposed to be covered by doctrine.

So what does this investigation of UGC tell us about the user? Essentially, we see that the contemporary user is rhetorically and legally tied to either the pirate or author in different ways, depending on particular contexts. For example, copyright doctrine tends to place the user closer to the pirate because so many uses are not legalized in copyright regimes. Law only supports particular uses, such as the use of copyrighted content for research or study, or in non-commercial contexts. Furthermore, the actions of users are often read as exceptions to be defended rather than positive rights to be asserted. This ideology is subsequently deployed in tools such as Content ID, which assumes that UGC is most likely infringing copyright when an uploaded video is matched against the content of a partner. In practice, however, the users indicated by the term UGC seem to hold a closer relationship to authorial subjects. Although they may be "using" copyrighted content, "Let's Play" YouTube stars do not feel that this should impinge on how their creativity is viewed or affect their ability to make money off of it. The development of social media economies has ultimately moved users toward an authorial subjectivity. As we will see in the next section, however, different technological contexts can shift the relational user back in the opposite direction, toward the much-maligned subject of the pirate.

Digital Innovation, Intermediaries, and the User: The User and Distribution

Over the last few years, US and Australian courts have had to come to grips with applications that challenge copyright law's existing approach to the broadcasting of film and television content. In the early 2010s, Optus (Australia's second largest telecommunications company) developed a service called TV Now, which allowed individuals to quickly and easily record free-to-air (FTA) television content and retrieve this content at a later date on a range of devices. At around the same time, Aereo, a now-defunct US technology company, released a service that allowed individuals to stream near-live broadcast television. In both cases, the companies were relatively confident in the legality of their applications when they went to market because they believed they had construed their services in such a way that they did not infringe on any exclusive rights.

Aereo's service rested on a reading of the US Copyright Act's "transmit clause," which prohibited anyone from communicating a performance or display of a work to the public "by means of any device or process, whether the members of the public capable of receiving the performance or display receive it in the same place or in separate places and at the same time or at different times."[66] Aereo avoided communicating to the public by placing thousands of antennas inside a Brooklyn warehouse that was strategically located to pick up FTA broadcast signals.[67] The service then sent these signals on to customers who tuned "into individual antennae to stream near-live TV to themselves, online."[68] Because Aereo used individual antennae for each transmission, it was able to produce single copies of a broadcast to each individual and thus purportedly avoid reproducing a copyrighted work to a public audience.

In a similar fashion, Optus's TV Now service was structured to rely on the Australian time-shifting provisions discussed earlier in this chapter to keep the service legal. Optus developed and launched TV Now on the basis that individuals would use the service in a similar way to a VCR. All a subscriber needed to do was search through an electronic program guide, press a record button, and then watch their show. It was clear in Optus's communications with customers that it had relied on these practices being viewed as private use to allow them to fall within the scope of the time-shifting exception. The company stated that "Optus TV Now is for your individual and personal use" and advised subscribers that it was "a breach of copyright to make a copy of a broadcast other than to record it for your private and domestic use by watching the material broadcast at a more convenient time."[69] But it was not just a service for storage. Like Aereo, the service effectively allowed the almost immediate rebroadcasting of live television as "playback could commence as few as two minutes after the start of [a] broadcast."[70]

These attempts to work at the edges of copyright law were quickly challenged by rights holders. They argued that Aereo and Optus were infringing copyright by essentially making copies of protected material through their applications and communicating this copyrighted material to the public. In the US, a consortium of large broadcast networks brought an action against Aereo,[71] whereas in Australia, it was Optus who initiated proceedings, accusing the Australian Football League (AFL) and the National Rugby League (NRL) of making groundless threats.[72] The AFL and NRL had contacted Optus because in addition to infringing their copyright, the TV Now's short playback time challenged the exclusivity of their broadcast rights agreements. Telstra (Optus's direct competitor) had recently paid the AFL AUD$153 million for the exclusive right to broadcast its games online.[73] If TV Now was declared legal, Telstra's exclusive rights would arguably be close to worthless. Therefore, the AFL and NRL, along with Telstra, made a counterclaim in response to Optus, arguing that TV Now was infringing their copyright.

At first glance, the user appears to be somewhat absent from the proceedings in both of these cases with rights holders facing off in court against an intermediary. Yet the question of use and the position of the user were both central to these debates because one of the few ways that both companies could maintain their legal status was to devolve responsibility to the user. The unique structural organization of the services formed the basis of both companies' arguments at trial. Optus argued that the user, rather than the company, was the primary actor copying television broadcasts. Because a user had to actively press record to make a copy, the copies made by the TV Now service should be protected by the time-shifting exception.[74] Aereo relied on an equivalent legal defense in the US, where time shifting for personal use is allowed under their broad fair use provision. Its position was that the technology and context in which this provision was deployed should not matter. Even if users were making copies by purchasing a service that made use of individual antennae residing in a Brooklyn warehouse, they were still time shifting for personal use.[75]

These arguments were initially successful, with both services found not guilty of copyright infringement. The Federal Court of Australia largely agreed with Optus's characterization of its service and found that the service fit squarely within the time-shifting provision and, indeed, the overall goals of Parliament when enacting that reform. Justice Steven Rares noted that the time-shifting provision was introduced with the intention of allowing an individual to

time-shift by making a copy of a broadcast that he or she could watch or listen to at a more convenient time. The TV Now service provides the user with a means for him or her to make a film of a broadcast.[76]

Subsequently, Justice Rares explained that because a user had to make an active decision to record a film, and "he or she alone did the acts involved in recording the copyright works," it was "impossible to say that Optus makes any of the films in the four formats that are created when a user clicks 'record' and its datacenter carries out that instruction."[77]

In the US, the consortium of major broadcasters originally moved for a preliminary injunction in the US District Court for the Southern District of New York, hoping to stop Aereo's activities. However, Aereo had knowingly structured its service on the back of a recent court case, *Cablevision*, whose ruling was that a remote storage system did not infringe the Copyright Act's public performance right.[78] Aereo was expecting that any case brought against it would fail because of this precedent and, initially, that is exactly what happened. In line with this previous case, Judge Alison Nathan summarized Aereo's argument as follows:

Aereo contends that, like the RS–DVR system in Cablevision, its system creates unique, user-requested copies that are transmitted only to the particular user that created them and, therefore, its performances are non-public ... each user is receiving a distinct transmission generated by their own individually rented antenna.[79]

Because the hearing was centered on the granting or denying of an injunction, the court did not need to decide whether Aereo or the user made the
copy or resolve the argument around Aereo's antennas.[80] Instead, the decision to deny an injunction was based on the precedent set by *Cablevision*.

Unsurprisingly, the respective rights holders appealed the decisions and
each case moved up the judicial hierarchy. In Australia, the Full Court of
the Federal Court of Australia heard the appeal and came to a different conclusion. The bench found that either Optus on its own or Optus in tandem
with users of the TV Now service made the copies, stating that "Optus'
role in the making of a copy ... is so pervasive that, even though entirely
automated, it cannot be disregarded when the 'person' who does the act of
copying is to be identified."[81] The moment it was decided that Optus had
a role in making the copies, the telecommunications company no longer
enjoyed the protection of the time-shifting provision. In its judgment, the
Full Court further clarified its interpretation of this provision, noting that
it was not meant to "cover commercial copying [conducted] on behalf of
individuals."[82] Optus sought to have the case heard in the High Court but
they were denied leave to appeal the decision.

In the US, Aereo survived the first appeal of the decision not to grant an
injunction, with the Court of Appeals for the Second Circuit supporting
the original District Court judgment.[83] In an interesting move, however,
both parties appealed to the Supreme Court to hear the case. The group of
broadcasters obviously did so with the goal of shutting Aereo down once
and for all, whereas Aereo wanted to get a final confirmation from the
highest court in the country that it had a legal right to operate.[84] Aereo was
understandably confident, having seen two separate courts confirm that
its service was legal. But this confidence was misplaced. In a 6–3 majority decision, the Supreme Court found that Aereo performed copyrighted
works publicly and was therefore infringing copyright. The majority held
that Aereo's service communicated to the public "because Aereo commu-
nicates the same contemporaneously perceptible images and sounds to a
large number of people who are unrelated and unknown to each other."[85]
The court attempted to severely restrict its decision to the performance of
copyrighted works, noting that it did not consider "whether the public performance right is infringed when the user of a service pays primarily for
something [like] the remote storage of content."[86] It is questionable how
successful this will be, though, because as the dissenting judgment notes, it

will now take "years" to decide which automated systems should "get the Aereo treatment" and which ones should be considered legal.[87]

Both of these cases radically altered the personal video recorder market, arguably making remote time shifting unlawful in Australia[88] and placing cloud services under legal uncertainty in the United States.[89] But what implications do these cases have for the user? The decisions of the Full Federal Court of Australia and the US Supreme Court reveal a reconsideration of use in both countries. In Australia, if a remote automated service makes a copy of a work it cannot be protected by personal use exceptions. In the US, it was found that individual acts of rebroadcasting still constitute a performance to the public, and so these acts become infringing and not subject to fair use protections. In both jurisdictions, a clear cultural decision has also been made: commercial companies should not broadly interpret exceptions created for users and expect to be protected as an intermediary on that basis. The judiciary only interpellates particular actors as users (rather than pirates) and it is clear that the practices Aereo and Optus engaged in did not fit with its expectations of what a "user" should look like.

In contrast, over the last decade or so the Supreme Court of Canada has afforded protection to commercial and non-commercial Canadian intermediaries under relevant fair dealing exceptions, offering an alternate conception of how an intermediary user can sit within a relational triad. In *CCH Canadian Ltd. v. Law Society of Upper Canada* (referred to in this chapter as *CCH*), a group of legal publishers accused the Law Society of Canada of infringing copyright in a range of legal materials by maintaining a custom photocopy service out of the Great Library in Osgoode Hall, Toronto.[90] The service allowed individuals to request particular articles, which would then be photocopied by library staff and sent directly to the requester. In addition to this, the plaintiffs also claimed that the Law Society's provision of self-service photocopiers was an authorization of copyright infringement.[91] The Supreme Court found that the Law Society did not authorize copyright infringement and argued that in relevant (but not binding) precedent set in Australia,[92] the library lacked "sufficient control over the Great Library's patrons to permit the conclusion that it sanctioned, approved or countenanced the infringement."[93] Just as notably, the court found that the remote photocopy service fell firmly within the scope of Canada's fair dealing exception for research or private study.[94] The fact that lawyers were using the service to support their commercial operations was not seen to be relevant and "[t]he reproduction of legal works" was found to be "an essential element of the legal research process."[95]

Helpfully, the Supreme Court of Canada articulated the wider implications of this decision when it heard *SOCAN v. Bell Canada* in 2012. The Society of Composers, Authors and Music Publishers of Canada (SOCAN) asked for royalties to be paid to the society when users of various online music services played previews of their members' music before making a purchasing decision. The Copyright Board and the Federal Court of Appeal both found that playing a preview of a song fell under the fair dealing exception for research and study and was not an infringement of copyright. The Supreme Court agreed and the majority judgment offered a detailed outline of a relational user, one that had a particularly close relationship with intermediaries who could assist individuals engaged in legitimate forms of use, suggesting an alternative vision of the intermediary–user relationship to that of the US and Australia. The court explained that the abstract concept of "the user" needed to be moved between different parties as the process unfolded, noting that "[t]he purpose of 'research' should be analyzed from the perspective of the consumer as the ultimate user, not the online service provider."[96]

The court refuted SOCAN's argument that "the purpose of 'research' should have been analyzed from the perspective of the online service provider," which would mean that "the purpose of the previews was not 'research,' but to sell permanent downloads of the musical works."[97] Instead, the court followed the precedent set by the decision in *CCH*, in which it "did not focus its inquiry on the library's perspective, but on that of the ultimate user, the lawyers, whose purpose was legal research."[98] In the same manner, the court explained that even though the service provider's use was commercial, "the ultimate users of the previews" used them to "help them research and identify musical works for online purchase," which fell under the fair dealing protections.[99]

Moreover, in *Alberta (Education) v. Canadian Copyright Licensing Agency (Access Copyright)*, which was also heard by the Supreme Court in 2012, the court directly acknowledged the sort of relational interactions that it expected to see between various users and intermediary bodies.[100] The Canadian Copyright Licensing Agency wanted teachers to pay for a license when they made copies "at [their] initiative with instructions to students that they read the material."[101] The court found that the teachers were engaging in fair dealing and explained that "[t]he teacher/copier shares a symbiotic purpose with the student/user who is engaging in research or private study."[102] Once again the court focused on the primary (or end) user rather than the intermediary and allowed the teachers to continue copying.

Across all three cases, the Supreme Court of Canada allowed a variety of intermediaries to participate in use as long as the final user was engaged in a legitimate use as defined by the fair dealing exceptions. This is a notable difference from the US and Australian cases in which the judiciary reacted strongly to the involvement of commercial intermediaries in the process of use. Indeed, the user articulated by the Supreme Court of Canada over a series of cases stands out as a highly relational one. The court expects the end user to be linked to a wider network of actors that engage in their own use in order to facilitate the specific use anticipated by fair dealing. This is an important legal construction of the user. It recognizes that when considering fair dealing exceptions, a user does not stand (or research and study) alone and so, subsequently, a defense under fair dealing may not be tied to just one user but could involve a chain of actors involved in a (legitimate) process of "use," all of whom should be acknowledged. This complex approach to use is particularly valuable because it leaves the legal door open for prospective technological innovation (as opposed to the decision regarding Optus) and offers a broad reading of fair dealing that supports the exception's central goals.

The different approaches that we have seen to both the intermediary and the question of use highlight once again how complicated the user is as a subject. Figure 4.2 shows where the various actors featured in this chapter as "users" sit across the relational triad. Relationality is present in law as attempted acts of use are reidentified as piracy or authorship, but it is also present in discourses around creativity and digital media where the user has been granted a series of authorial capacities from the mid-2000s onward (as noted in chapter 2). So the question posed at the beginning of this chapter arises again: *who* is a user? Should intermediaries (and, in particular, commercial intermediaries) be incorporated in this subject position? How much latitude should they be granted in structuring their operations in relation to existing user exceptions? And how do we meaningfully distinguish between alternate modes of authorship at a time when people are increasingly consuming commercially viable user-generated content?

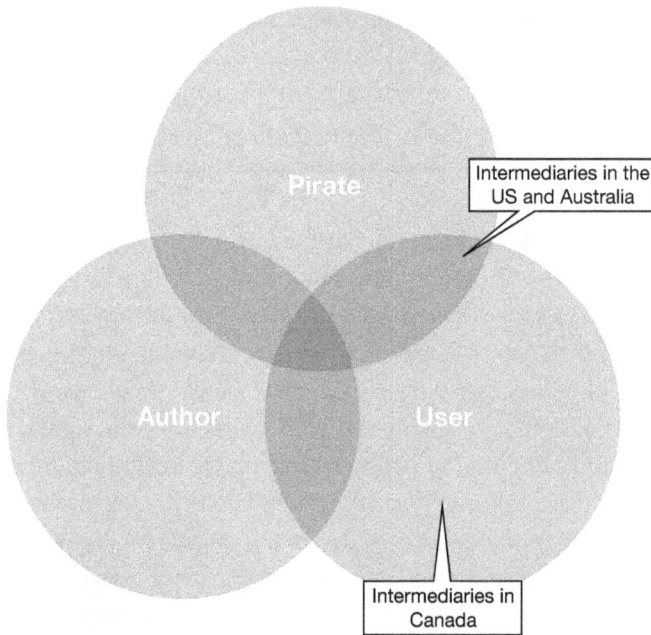

Figure 4.2
Canada views intermediaries as users, whereas in recent cases the United States and Australia have viewed intermediaries as users who engage in piratical actions.

In light of these questions, I will end the chapter by discussing what sort of scholarly inspiration law and Internet studies can draw on in order to more accurately locate and understand this figure.

A Way Forward

As noted earlier, legal scholars have begun to develop a clearer understanding of the user. Cohen identifies three main models of the user that have become prominent in discussions around copyright law:

[T]he economic user, who enters the market with a given set of tastes in search of the best deal; the "postmodern" user, who exercises limited and vaguely opposi-tional agency in a world in which all meaning is uncertain and all knowledge rela-tive; and the romantic user, whose life is an endless cycle of sophisticated debates about current events, discerning quests for the most freedom-enhancing media technologies, and home production of high-quality music, movies, and open-source software.[103]

She suggests that each model is "more unrealistic than the last" because "none of them provide ... a convincing model of how real users behave."[104] As an alternative way forward, Cohen offers her concept of the situated user, a subject that "appropriates cultural goods found within her immediate environment for four primary purposes: consumption, communication, self-development, and creative play."[105]

This formulation, which Cohen develops in later work, significantly and productively repositions the user in relation to copyright. Instead of relying on the "conventional dichotomies between author and consumer, author and imitator, author and improver, and author and critic that pervade the copyright literature,"[106] we can view the situated user as a subject embedded in an interconnected and relational network of actors, resources, and emergent creative practices. As already noted, Papparlado has offered the most substantial theorization of the user in law, extending on Cohen's conceptual work.[107] She calls for an understanding of the user as an autonomous subject with meaningful interests in self-expression, establishing personal connections, becoming educated, participating in communities, and engaging in play.[108] With this concept in mind, Pappalardo contends that copyright exceptions should be interpreted broadly to support passive consumption, active consumption, cultural engagement, and ongoing creative acts.[109]

Still, there is a need to move beyond the necessary abstraction that law and legal scholarship engages in because the user is also constituted through the quotidian legal understanding established through industry discourse and interpretations of law in the broader community. Therefore, as Cohen notes, it is not enough to simply understand the user abstractly; we must delve into the messy and conflicting desires of the various actors that operate under this subject position.[110] This is also a task for scholars of digital media, who, as noted earlier, likewise use the term as an easy shorthand for a variety of actors with various motivations.[111]

One way forward is to take note of disciplines such as STS, which has begun to tackle the problem of the user in more depth than either law or media studies. This discipline has suffered from similar problems, with Maria Bakardjieva explaining that when it comes to technology, "the user is a marginal figure ... as a subject, but has a central place in it as an object."[112] Therefore, much like we have witnessed with regard to copyright law, computing also sees an array of powerful actors coming together in an attempt to define the user from above. Steve Woolgar conducted a study of how a microcomputer product (only ever referred to by the acronym DNS) was designed and noted that:

In configuring the user, the architects of DNS, its hardware engineers, product engineers, project managers, salespersons, technical support, purchasing, finance and control, legal personnel and the rest are both contributing to a definition of the reader of their text and establishing parameters for readers' actions. Indeed, the whole history of the DNS project can be construed as a struggle to configure (that is, to define, enable and constrain) the user. These different groups and individuals at different times offered varying accounts of "what the user is like."[113]

In these situations, the user of technology is imagined and reimagined by various actors, regardless of the fact that "[u]sers are hard to perceive as a social group ... because of their dispersed state of existence [and] their diverse cognitive and material resources, interests and ideologies."[114]

Therefore, STS has turned to detailed ethnographic studies in order to explore how users are constituted in relation to technology. Through this research scholars have noted that many processes of technological design locate the user in a position of passivity.[115] This is interesting not just because of its similarities to how law sometimes constructs the user, but also because of the recognition that the user carries this passivity as a cultural marker. The cultural associations around users, authors, and pirates do not necessarily go away when we transplant these terms to law or to discussions around digital media more generally; indeed, they are more likely to be reinforced. Much like the group input during technological development identified by Woolgar,[116] when it comes to the types of user-oriented law reform discussed in this chapter, the debates tend to feature lawyers and representatives from authors' groups and other rights holders, government, and even user rights groups. But we rarely hear from everyday users themselves. As with the development of DNS, the user is defined, enabled, and/ or constrained by an array of expert actors that may or may not have their interests at heart. This means that ultimately, in both law and the design of technology, the user becomes "embodied in the machine" (or the law), and it is only after the key decisions are made that they "are confronted by the preconceptions of themselves reflected in the design" of law or digital technologies.[117]

This discussion about science and technology studies is not merely instructive. Although this discipline points toward ways of grasping the subject of the user in a more detailed fashion, its central concern—the relationship between individuals and technology—is now also the concern of copyright. The user in copyright law is now socio-technical. As I noted earlier, many methods of copyright enforcement from TPMs and DRM to YouTube's Content ID system have structured technological systems around particular imaginaries of uses and users. This means that the content user is often already ensconced in a technological relationship, and so technology as well as copyright law forms part of this subjectification. Therefore, it is worth thinking about these studies from STS in future explorations of how law, formal and informal legal cultures, and technology imagine users. Considering the unstable nature of the relational user identified in this chapter, a detailed history of this subject and the ongoing study of how it is constituted in law and culture are both necessary scholarly tasks.

5   Reimagining the Pirate: Approaching Infringement Relationally

This chapter proceeds from the premise that it is worthwhile to examine the socio-legal discourses that surround the pirate. This subject tends to be constituted through a variety of competing rhetorical claims rather than in reference to the minutiae of copyright law.[1] I will explore these different rhetorical claims and also examine what happens when non-legal actors resist particular modes of subjectification in favor of others. There is value in analyzing these cultural discourses as they can shed light on how power is constituted in law and also tell us more about how the process of subjectification operates within the context of our relational approach. The chapter also examines the links between infringement and authorship through an exploration of creation in low IP (intellectual property) environments and copying practices in media industries and on social media. The analysis provides further detail around processes of subjectification in which some copiers are designated pirates and others authors, and draws attention to the links between copying and modes of authorship.

The chapter begins by arguing that the pirate subject can be productively examined beyond the confines of legal doctrine, before moving on to a study of the protests against the Stop Online Piracy Act (SOPA)[2] and PROTECT IP Act (PIPA),[3] which occurred in 2012. The discourse that emerges from these protests reveals clear interrelations between the pirate and the user, shedding light on how Western piracy is recovered from the edges of law. The chapter then examines the suppressed relationship between the author and the pirate by outlining the productive elements of infringement. We see that copying can sustain creativity, enhance communities on social media platforms, and function as a driver of economic activity.[4] Collectively, this approach recuperates the pirate and excavates the residual subjects of the user and author that emerge and disappear at different times during these debates.

Identifying the Pirate: Moving Beyond Law

There is real value in exploring and analyzing the pirate beyond the boundaries of legal doctrine. Over the last decade or so, a number of scholars have productively critiqued the strict application of copyright law. Lawrence Lessig tells the now relatively well-known story of Stephanie Lenz, who filmed her eighteen-month-old son dancing to Prince's "Let's Go Crazy" and uploaded it to YouTube.[5] Universal Music Group eventually found the video and demanded that YouTube take it down, and the platform complied. Lessig argues that this sort of amateur creativity should be allowed to flourish online and be left "free from regulation."[6] Anupam Chander and Madhavi Sunder make similar arguments with respect to fan fiction, calling for these transformative creative practices to fall under the fair use exception and suggesting that "authors should not readily 'cease and desist,' as copyright owners demand."[7]

These studies examine how particular forms of creativity are framed by narrow interpretations of copyright law that attempt to position them as infringing. They also usefully discuss the cultural implications of the over-enforcement of intellectual property rights, and the links between cultural development and copyright. Yet, law still stands as the central organizing principle of these analyses, which results in piracy being read (and the "issue" ultimately being solved) through a legal lens. This involves a familiar analytic process. First, an analysis of the problems and contradictions of the legal framework takes place, and then through locating unjust or illogical applications of the law, a critique and diagnosis of intellectual property law can occur. Scholars aim to recuperate "everyday," formerly legal practices that have since been made illegal, and a call for "balance" and a prescription for legal reform is offered.

It is worth noting which practices are not accounted for in these analyses. Many scholars are willing to argue for the recuperation of transformative amateur creativity or even minor instances of file sharing, but they often distance themselves from or seriously qualify their support for commercial piracy. Lessig argues that commercial piracy "is not just a moral wrong, but a legal wrong."[8] Similarly, Lucas Hilderbrand is willing to recover bootlegging from its "negative" or "criminal" connotations but he refuses to acknowledge more substantial acts of piracy.[9] For Hilderbrand, bootlegging is concerned with the egalitarian or productive redistribution of culture and information, whereas pirates steal for profit.[10]

This restrictive approach places law at the heart of what is essentially a cultural conversation. Piracy is not just an offshoot of law but a complex social and cultural act carrying a set of broader agencies beyond that of "infringement." Adrian Johns supports this sociocultural approach and notes that historically, piracy has been more than just "a mere accessory to the development of legal doctrine":

Piracy cannot be adequately described, let alone explained, as a mere byproduct of such doctrines. It is empirically true that the law of what we now call intellectual property has often lagged behind piratical practices, and indeed that virtually all its central principles, such as copyright, were developed in response to piracy. To assume that piracy merely derives from legal doctrine is to get the history—and therefore the politics, and much else besides—back to front.[11]

Although Johns is speaking about historical analysis, his critique still holds when analyzing piracy and the pirate today. There is an inherently close relationship between piracy and copyright doctrine, of course, but the subject is as much culturally as it is legally determined.

A diverse group of scholars have approached piracy as a productive cultural force, with only passing reference to the legal doctrine that ostensibly seeks to regulate the practice. As noted in the introduction, Rebecca Tushnet offers an interesting challenge to the embedded logic of copyright law, arguing that particular forms of non-transformative copying can serve the general social good by functioning as free speech.[12] Referencing the First Amendment to the US Constitution, she argues that "copying may sometimes be an instance of free speech even when it is also copyright infringement" and details how copying can assist self-expression, persuasion, and affirmation (for example, when reciting the Pledge of Allegiance).[13] Tushnet still makes a distinction in regard to commercial piracy, arguing that there is "no free speech right to download entire works for which [she] could readily pay," but her analysis is refreshing for the way it approaches copying in a more holistic fashion, moving beyond the fixation on transformation that the fair use exception occasionally engenders.[14]

Cultural research takes this critique even further by drawing extensively on empirical research and cultural theory. Kavita Philip contends that the pirate has largely "functioned as a raced, gendered subaltern" and points out that calls for copyright reform in the West not only identify the pirate as a "limit point" but tend to locate this limit point in Asia, which stands as a representation of "bad piracy."[15] This allows copyright activists in the West to support the development of new technologies like peer-to-peer (P2P) networks or remixing while firmly situating this support within the confines of "western liberal democratic law."[16] Small-scale Western copyright infringement can thus be recuperated as an inherently democratic and non-threatening practice. Lawrence Liang furthers this critique by exploring how piracy operates in relation to the purchasing power of citizens from particular countries. His analysis starts from a similar place as Philip's, asking a pertinent question: Why can an illegal practice like downloading music "find redemption," whereas other forms of piracy cannot?[17] He goes on to suggest that the piracy of cultural products in the Global South is "tainted" by commerce, making it harder to justify than infringing intellectual property in order to access medicines or learning materials.[18]

Over the last decade the pirate has also emerged as a generative political subject and a number of media and cultural studies scholars have analyzed the emergence of pirate parties in some depth. Martin Fredriksson has conducted sustained empirical research on these parties and discovered that they have concerns about a perceived loss of democracy due to the influence of corporations in political decision-making.[19] They also have a strong belief that the political contestations around today's information society are essentially questions about civil liberties, and this has become central to the platforms of various national pirate parties.[20] This turn to the political is unsurprising if we consider the fact that a number of copyright activists (including, most notably, Lawrence Lessig)[21] believe that fairer copyright laws are essentially unachievable due to the tight structural connections between political representatives and major content industries, formed through established lobbying networks and political donations. Still, as Patrick Burkart notes, despite brandishing the representational figure of the pirate, these parties represent a somewhat minor form of middle-class activism rather than a serious call for revolution.[22]

The rest of this chapter will be dedicated to further exploring these discourses around the pirate, identifying when authorship and use are evoked or ignored in these emotive debates, and advocating for a reassessment of the pirate in general. We see that our current understandings of the subject are tied to particular cultural views around the proper use of existing corporate networks of production and consumption. Citizens of countries in the Global North regularly call on the subject of the user to disavow or dismiss their piracy and, as Philip notes, locate piracy overseas.[23] Nonetheless, further examination of how piracy operates in various industries, on social media, and in the Global South ultimately leads us to an understanding of the pirate as a generative subject. The cultural capacities and affordances of authorship, whether they involve claims of creativity, innovation, or cultural or economic progress, are not unique to the author.

SOPA and PIPA: Middle-Class Users and the "Foreign" Pirate

SOPA and PIPA were two infamous antipiracy bills introduced to the US House and Senate with a significant amount of fanfare in 2011. The Democratic and Republican parties supported the bills and the Motion Picture Association of America, the Recording Industry Association of America, and the US Chamber of Commerce also publicly backed the reforms.[24] The bills focused on copyright infringement occurring on offshore websites and introduced a range of new powers for copyright enforcement. Under SOPA, the US Attorney General could demand that a website be removed from the Internet if it offered goods or services that engaged in, enabled, or facilitated copyright infringement.[25] Another provision in the bill stated that Internet service providers (ISPs) would have to "take technically feasible and reasonable measures designed to prevent access by its subscribers located within the United States to the foreign infringing site," which suggested that ISPs would have to monitor subscriber activity in some fashion.[26] PIPA offered an alternative way to stem copyright infringement by requiring "domain name system providers, financial companies, and ad networks" that held a relationship with copyright infringing websites to alter their practices to avoid making copyright infringing material easy to access.[27] The bill would also allow the Attorney General to sue the owner or operator of an infringing website.[28]

Although these legislative proposals were welcomed by politicians of all stripes as well as lobbying organizations for various content industries, they were strongly opposed by many legal experts, Internet activists, and major players in the tech industry. The opposition's major concern was the wide scope of liability that SOPA and PIPA established around copyright infringement, along with the proposed deployment of a range of highly controversial enforcement mechanisms. Edward Lee notes that the strategies PIPA authorized, namely, the ability of the US Attorney General to make foreign websites effectively disappear from US search engines, bore an incredible similarity to the practices used by China to enforce its contentious online censorship regime, known as the "Great Firewall of China."[29] The bill also demanded that tenuously connected intermediaries (such as financial companies and ad networks) become involved in copyright enforcement efforts, which was not well received. If enacted, the bill would have required these parties to take measures to avoid linking to (or working with) infringing overseas sites, shifting the burden of enforcement directly onto major transnational companies like PayPal and Google.

Criticism of SOPA was even more damning. The major problem with the bill was its broad remit. It was structured in such a way that US citizens would be unable to access a website if it offered goods or services that engaged in, enabled, or facilitated copyright infringement. Because the bill focused on actors that assisted infringement as well as infringing conduct, a broad reading of this provision would make much of the Internet in breach of the proposed law. As Michael Carrier noted,

[a]ny means making it easier for others to access copyrighted content could be punished. Such a standard could ensnare in its grasp numerous websites and services, including YouTube, Google, Facebook, Flickr, Dropbox, and blogs, each of which could be found to enable or facilitate infringement.[30]

The expanded danger of liability represented a direct threat to a group of highly successful and previously "lawful U.S. Internet companies," and so the tech industry was unsurprisingly worried about the bill.[31]

The bills' most vociferous critics were initially a familiar group of actors, namely, digital rights activists, nonprofits, and academics. Internet engineers and law professors were signing letters and sending them to Washington but the looming threat of SOPA and PIPA had not yet captured the imagination of the general public.[32] Furthermore, although the tech industry was worried about the bill, its concerns were not yet made public. Indeed, it was only thanks to the efforts of the digital rights nonprofit Fight for the Future that the public at large was made aware of the potential threats of SOPA and PIPA. As Lee explains, the group was able to spread the word by arguing that the bill essentially introduced a form of online censorship "because it allowed the Attorney General to blacklist domain names and make the content of the affected websites effectively disappear—without adequate safeguards to ensure the site was criminally infringing or deserved to be shut down."[33]

Fight for the Future eventually contacted the wider tech industry through the Mozilla Foundation, the organization that developed the open-source browser Firefox, and "informed the group that, not only did SOPA have a chance of passing in the House, *the bill will pass.*"[34] The threat of the bill passing with little debate immediately energized the industry and discussion turned to the best way of protesting the bill and raising awareness about the issue. One of the Fight for the Future members suggested that since they were organizing the campaign around censorship, one way to protest the bill would be for sites to artificially "censor" themselves to show what would happen if SOPA passed. This was a compelling tactic but raised issues for some of the larger organizations in the room like Google, which had never "blacked out its site and hadn't ever used its home page in such a political way."[35]

There was extensive discussion about whether an online protest directed at the general public would be a good use of time and resources. A number of prominent parties including Google withdrew their support for a blackout, but eventually Fight for the Future received support from Tumblr and a highly successful day of online protest occurred on November 16, 2011.[36] SOPA was not withdrawn (although some amendments were made) but the protest was successful in mobilizing a wider group of concerned Internet users, as well as additional members of the tech industry, in a public campaign against the bill. So, Fight for the Future, along with its new coterie of tech companies, aimed for a second online blackout on January 18, 2012. Importantly, Google and *Wikipedia*, two of the largest sites in the world, agreed to participate, as did more bespoke online communities such as Tumblr and Reddit. The involvement of Google and *Wikipedia* dramatically extended the impact of the protest, with 162 million people visiting *Wikipedia* during the blackout only to be redirected to information about SOPA and PIPA (see figure 5.1).[37]

Alongside these online protests, a modest number of people took to the streets in New York, San Francisco, and Seattle (see figure 5.2). Significantly more people tried to contact their representatives directly to voice their concerns, to the extent that

Engine Advocacy, a service that helps people call their local members of Congress, said that as many as 2,000 a second were trying—demand so heavy that many of the calls could not be completed.[38]

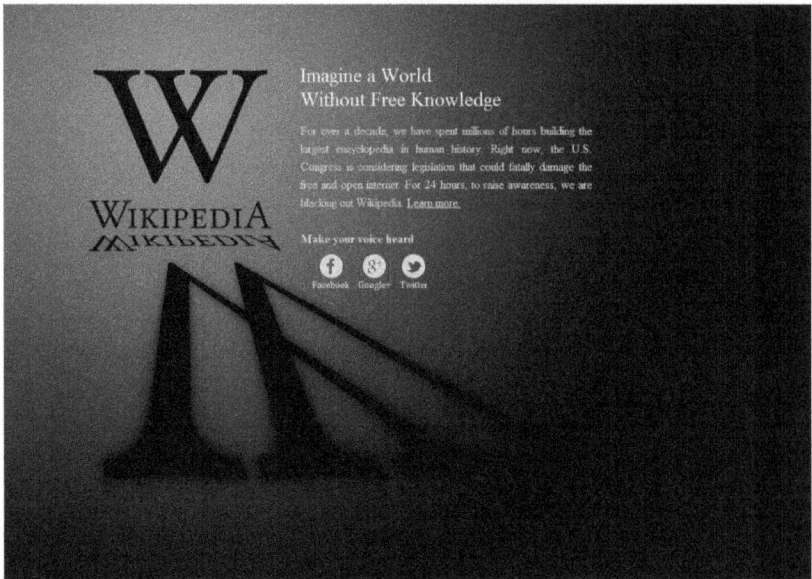

Figure 5.1
Wikipedia was one of the websites that participated in the blackout protest against SOPA and PIPA. Credit: David Holmes (CC BY 2.0).

It took only a few hours of this global protest before lawmakers started publicly withdrawing their support for the bills, and PIPA and SOPA were pulled from consideration two days after the January protest.[39] The story of these protests is compelling: a small nonprofit mobilizing major tech companies and eventually much of the online community to take down two pieces of legislation that had received support from all the key players in Washington. But there is another story to be told here about the different ways the bills and the protest movement positioned piracy and the pirate.

Both SOPA and PIPA replicated the Western-centric discourse around piracy discussed earlier in this chapter. This was particularly apparent after the first round of protests. SOPA's sponsor Lamar Smith argued that the bill targeted "rogue sites" that functioned beyond US borders:

These foreign websites are called "rogue sites" because they are out of reach of U.S. laws. Movies and music are not the only stolen products that are offered by rogue sites. Counterfeit medicine, automotive parts and even baby food are a big part of the counterfeiting business, and pose a serious threat to the health of American consumers.[40]

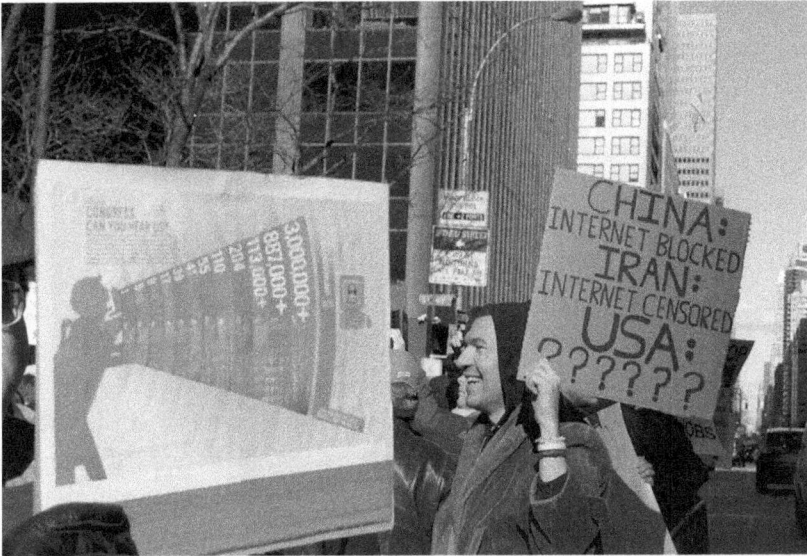

Figure 5.2
Protesters on the street in New York. Credit: Guy Dickinson (CC BY-SA 2.0).

In a similar fashion, PIPA explicitly sought to enhance enforcement measures "against rogue websites operated and registered overseas," and this is how the bills were primarily sold to the American public.[41] As the quote above reveals, however, they were also positioned as a way of protecting the American consumer. Online copyright infringement was regularly occurring within the United States but "piracy" and the pirate were located overseas, with the American citizen framed as a law-abiding consumer.

The SOPA and PIPA protests "located" the pirate in a similar fashion. For example, one of Google's public statements on SOPA criticized its attempt to "censor the Internet" but did suggest that there were "smart, targeted ways to shut down foreign rogue websites."[42] This statement recognized the pirate as a subject that was both present in these public policy debates but also firmly located outside of America's borders.

There was also an attempt to link criticisms of the bills to broader concerns around free speech and freedom more generally. The statement accompanying *Wikipedia*'s decision to blackout its front page captured the tenor of many of these protests, with the Wikimedia Foundation Executive Director Sue Gardener stating that the bills were a direct threat to the "free and open" web and voicing *Wikipedia*'s support of "everyone's right to freedom of thought and freedom of expression."[43] This narrative was echoed on the busy *Wikipedia* page "Wikipedia:SOPA Initiative/Action," where individuals debated whether to black out the site. Contributors argued strongly in favor of the blackout, claiming that SOPA would "destroy our freedom, our internet, our digital frontier" and "prevent global free speech," and that the "potential consequences of this bill on the internet and free speech [were] dire indeed."[44]

This discussion around freedom and free speech pointed to another notable facet of the SOPA/PIPA protests: protesters fought these bills by adopting the subjectivity of the user. Instead of specifically mentioning piracy or relating to the subject of the pirate, they talked about the right to use an uncensored Internet and the need to protect online platforms like *Wikipedia* and Google. In a similar fashion, the ongoing mentions of freedom and the defense of free speech throughout these protests spoke to the "respectable face of middle-class copyright critique."[45] Despite the fact that copyright infringement was occurring in the United States at the time, the political response of protestors was not to recognize this piracy but to resist this subjectification and instead present as misidentified users who were bearing the brunt of an ill-informed and poorly executed antipiracy operation.

The preceding critique of SOPA and PIPA reveals the relational nature of the pirate in the contemporary West. Although the Pirate Party is a visible actor through its various national and local iterations, more often than not copyright infringement in places like the United States is defended or reidentified as use. This happens at a scholarly level with acts recuperated as either innovative (such as sampling) or defensible (for example, peer-to-peer file sharing), and it also happens at the level of public protest. The fact that foreign piracy is occurring overseas and should be stopped was never seriously questioned during the SOPA/PIPA protests, only the method of enforcement. This embrace of the user subject position and relegation of the pirate to foreigners comes to define citizens in the United States and other Western countries. They have been called pirates by industry groups like the Motion Picture Association and the International Federation of the Phonographic Industry throughout the 2000s and 2010s, but individuals can and often do resist this mode of subjectification and mobilize other socio-legal discourses to prevent being defined as unrepentant pirates. There is always the (comparatively) more authorized subject of the user available for them to inhabit. I will continue to discuss the relational connections between the Western middle-class user and the pirate in more detail in the next chapter. At this point, it is enough to point out that there is a close relationship between these two parties and move on to explore the interrelations that can occur between the pirate and the author.

The "Pirate" as a Productive Actor

There is a common presumption among many lobbying groups, legislators, and members of the general public that copyright is a vitally necessary legal technology for any innovative sector populated by creative practitioners. Yet a number of highly creative fields are either not protected by copyright or essentially ignore copyright laws that do exist for their supposed protection. Fashion designers,[46] comedians,[47] chefs,[48] and adult video producers[49] all operate in low IP or no IP environments. But despite the absence of rigorous and enforceable copyright, these sectors are still able to innovate. This is because in many of these industries, copying is good for business. Kal Raustiala and Christopher Sprigman have conducted an extensive overview of this phenomenon in their book *The Knockoff Economy* and suggest that a lack of IP protection can be beneficial. They note that in the fashion industry copying speeds up the fashion cycle, which means that clothing comes in and out of style more quickly. As popular designs get copied, they lose "distinctiveness," forcing leading designers to innovate once more, starting "new trends—and, as a consequence new sales."[50]

We also see copying emerge as a central driver of creativity in the arena of the professional kitchen. Neither recipes nor finished dishes are protected by copyright. Although some recipes can be "valuable business information"[51] and protected by trade secret law, for example, the recipe for Coca-Cola or Kentucky Fried Chicken's eleven secret herbs and spices, most recipes (even in high-end restaurants) can be copied in their entirety. The preponderance of copying is further entrenched in the restaurant business by the practice of "staging," in which "a young chef works essentially as an intern under the tutelage of another chef."[52] This method of instruction in fact *requires* the young chef to copy new dishes and techniques to build a repertoire of skills they can take with them after leaving. Copying functions as a necessary educational tool, which in turn generates innovation in the entire sector. The ability to copy both directly and indirectly (through recipes) allows chefs "to learn from one another, and thereby keep incrementally improving their offerings."[53]

Although chefs and fashion designers operate in industries rife with copying, they are nonetheless both professions that produce culturally (and occasionally, legally, in the case of fashion design) recognized forms of authorship. Thomas Keller and Giorgio Armani are seen as artists on a level with other cultural producers, from painters to composers, who produce work protected by copyright law. The most famous practitioners all have a style attributable to them and are original and innovative in their work, all key markers of authorship. Yet they are also all pirates who have copied and will no doubt continue to copy throughout their careers. This places the phenomenon of copying in an interesting position.[54] In the above cases, copying is not considered the obverse of authorship, but is seen as central to the ongoing maintenance of each industry. It functions as an educational tool, a marker of status, and a driver of innovation. Since copyright law and other IP regimes are largely not relevant in these industries, we cannot automatically call this infringement in the formal sense; instead, we can identify an alternate history of copying that functions as a productive force, as well as locate a blurring of the boundaries between authorship and piracy.

This overlap of the author and pirate also occurs on social media. The replicable nature of online content is one of the more noticeable changes of the last decade or so. Copying is now a standard practice on social media platforms, with individuals passing on videos of cute puppies, poignant photographs, relevant GIFs, and hilarious memes to friends, family members, and distant acquaintances.[55] Much of this material is copyrighted content, but the law is rarely enforced in these spaces. This is partly due to a discursive shift in how this distribution is understood. As Jenny Kennedy notes, "a technologically situated discourse of sharing has persisted throughout networked culture."[56] This is clearly identifiable in the redefinition of online copyright infringement as "file sharing,"[57] but also more recently in the invocations to share (rather than, say, copy) on social media platforms.[58] Sharing offers a positive way of thinking about this method of distribution (see figure 5.3).

Social media platforms have adopted this discourse of sharing as part of a broader strategy of engagement, with the goal of becoming economically viable businesses.[59] Numerous creators have also adopted this discourse and encouraged copying to support the wide distribution of their

I'm no longer allowed to teach Kindergarten.

Figure 5.3
"Sharing" is a contested term and often used in place of "piracy" in discussions about online distribution. Credit: Zach Weinersmith, smbc-comics.com/comic/sharing.

content. Part of the reason photographer Brandon Stanton (who runs the *Humans of New York* blog) developed such a substantial following is that he regularly posted his content on a Facebook page, allowing his photos to be shared across the platform. This is a common strategy nowadays, with numerous amateurs (and aspiring professionals)—from make-up artists to powerlifters—using the affordances of social media platforms to build an audience and, ideally, monetize their content.[60] Professional media organizations have also responded to this compulsion to share, structurally integrating with social media platforms to allow their content to be widely circulated. This changed mode of distribution and consumption has been most clearly described by Henry Jenkins, Sam Ford, and Joshua Green, who suggest that media is only able to survives if it is "spreadable" and able to be distributed across multiple platforms by both authorized and unauthorized actors.[61]

These changes in distribution suggest that much as it does for chefs and fashion designers, copying also structures the media industry and functions as a central site of economic generation for both large companies and small-scale creative entrepreneurs. These actors have numerous reasons for encouraging the sharing of their content and not enforcing potential claims of copyright infringement. For large media companies, this practice echoes the earlier broadcast media practice of providing content for "free" and generating income from advertising. Even if these companies ultimately demand payment for content at some stage through a paywall, allowing some content to spread online is a simple way of building an audience around a brand. This is also one of the primary reasons why individual entrepreneurs encourage the sharing of their content. Indeed, the YouTube Creator Academy encourages creators to "[s]hare [their] videos on linked social platforms" to "reach off-site audience members"[62] and "reach beyond YouTube."[63]

The intertwined phenomena of sharing and spreadable media also produce an authorial pirate. As noted earlier, part of the task of this chapter is not simply to identify piratical behavior that can be legally identified as authorship but also to locate the pirate as a generative actor that can draw on authorial capacities in a broad sense. The preceding discussion locates acts of infringement that are either tacitly accepted or encouraged by copyright holders to enhance their economic activities.[64] Therefore, the wholesale copying (or sharing) of copyrighted content can at times act as a generative economic practice. Thinking about creativity and production in a cultural sense then (rather than in terms of the letter of the law), the pirate no longer stands on the outer edge of copyright law but alongside the author as an economically and creatively productive subject.

We are not just limited to this sort of meta-level conceptual thinking to locate a relational connection between the pirate and the author. The spreadable nature of media has presented some interesting case studies in reuse and appropriation that clearly trouble the boundaries between infringement and authorship.[65] One of the most notable recent cases is that of Richard Prince, an appropriation artist who exhibited a number of other people's Instagram posts during the Frieze Art Fair. The photos were "blown up and jet-printed on six-foot canvas" but otherwise edited only slightly:[66]

Although he did not alter the usernames or the photos themselves, he removed captions. He then added odd comments on each photo, such as "DVD workshops. Button down. I fit in one leg now. Will it work? Leap of faith" from the account "richardprince1234."[67]

The prints were exhibited with little comment in 2014 at the Gagosian Gallery in New York, but when Prince sold most of the prints for US$90,000 each at the Frieze Art Fair in 2015, media outrage quickly followed.[68]

The media narrative centered on Prince's theft of the images, stating that an artist was making money from "stolen" photos and claiming that he was "controversial"[69] or, more bluntly, simply a "jerk."[70] The subjects of the photos themselves had a more complicated response to the practice. Unsurprisingly, a number of people were furious that their photos (and their identities) had not only been made public but also commoditized. An Australian photographer Peter Coulson felt "violated" and compared the act to having "your house broken into."[71] Prince had used one of his photos of a model, Alice Kelson, in his exhibition. Coulson had given Kelson permission to post his photo to Instagram, which is how Prince got access to it. Anna Collins, a college student, was similarly unimpressed with "a middle-aged white man making a huge profit off of [her] image."[72] It is worth noting, however, that "the shot of Collins and her boyfriend gazing at the glowing screen of a sticker-covered laptop" was actually taken by her sister, who would hold copyright in the image.[73] In contrast, Karley Sciortino, another photographic subject, felt honored to be included in a piece of Prince's artwork.[74]

Interestingly, a significant number of the photos chosen for Prince's exhibition were of models from the "adult lifestyle brand" Suicide Girls. The company's models are defined by an "alternative" look that incorporates tattoos, piercings, and dyed hair and are regularly featured on the brand's Instagram account. The brand took a novel approach to Prince's appropriation. Its founder Selena Mooney noted in an Instagram post that their images had often being stolen or reappropriated by other companies but felt that Prince's activities were different:

If I had a nickel for every time someone used our images without our permission in a commercial endeavor I'd be able to spend $90,000 on art. I was once really annoyed by Forever 21 selling shirts with our slightly altered images on them, but an Artist? Richard Prince is an artist and he found the images we and our girls publish on Instagram as representative of something worth commenting on, part of the zeitgeist, I guess? Thanks Richard![75]

But Prince did not get off scot-free. Though generous in her response to his work, Mooney also used the post to launch the Suicide Girls' new series of prints. The company would print any of the altered Suicide Girls images to the exact specifications of Prince's artwork at the discount price of US$90, which Mooney cheekily noted was "99.9% off."[76] Any profits would go to the Electronic Frontier Foundation. Prince publicly approved this response to his artistic practice on his Twitter feed.[77] There is a legal argument to be made about whether or not Prince's appropriation was fair use, but the more interesting issue emerges from the relationship between attribution and genre, a discussion begun in chapter 3. Just like the middle-class music downloader earlier in this chapter, artists are not always seen as pirates when they copy.[78]

The practice of appropriation also underlines the awkward tensions that emerge between the author and the pirate around the question of power. Collins, the college student whose image was featured in the exhibition, referenced the gendered history of authorship that prioritizes men over women. This history implicitly authorizes Prince's appropriation practices and allows him to make a profit from it. Indeed, in each of the examples of copying discussed above, a designation of authorship is as much the product of an existing system of power relations, and conventions of genre and style, as formal law. This of course reveals a stark division between authorship and piracy. Authorship is a mode of cultural authority often granted to those in power, whereas an individual is more likely to be aligned with the pirate if they lack power. Yet Prince's case also underlines the close but often obfuscated connections between the author and the pirate. As his activities show, these subjects are connected through many instances of actual creative practice, and it is often only through the direct or indirect deployment of power that boundaries can be established between these two figures.

As the discussion at the start of this chapter showed, this deployment of power is also present in global conceptualizations of piracy, where piracy in the Global South is starkly separated from Western practices of "sharing" or "remix." But this piracy can similarly be reframed as a generative activity, like the media practices examined above. Numerous scholars have noted that piracy develops "provisional and informal infrastructure," supporting new distributive pathways and forms of access for countries that have been left out of the institutional networks of modernity.[79] Joe Karaganis explains the everyday inequalities of distribution in these countries. Whereas "wealthy elites" in capital cities have access to "[m]ovie theatres, DVD and CD retailers, bookstores, and software vendors," "[s]maller cities and the provinces are chronically underserved."[80] Therefore, pirate networks allow media to flow to these areas, "radicalizing media access for subaltern groups" and "allowing the entry of vast numbers of poor urban residents into media culture."[81] In this context, piracy is not a parasitical act but rather a foundational prerequisite for media circulation that can fulfill a number of central social and cultural functions.[82]

Piracy is also related to pricing. Liang and Achal Prabhala note that citizens in the Global South have to "commit significantly higher proportions of their income" to purchase books.[83] Their comparative analysis of the cost of books in different countries is particularly illuminating, showing that the standard price of books in India is prohibitive for a large segment of the population:

It is instructive then that the prospect of paying $440.50 for Roy's *God of Small Things* in the USA is evidently alarming. Yet, the notion of paying $6.60 for the book in India (which in Indian terms is exactly the same value as US$440.50 in the United States, by this logic) is not treated with similar alarm.[84]

Therefore, piracy does not only function as distributive infrastructure or a site for consumption and pleasure, but also offers critical cultural infrastructure. This has had an impact in India, with piracy allowing "groups excluded from the technical education common in the Indian middle and upper classes" to "climb the value chain in the information economy."[85]

Another salutary aspect of generative piracy is its archival capacity. In his studies of pirate DVD distribution in Mexico City, Ramon Lobato has noted that some pirate stalls do more than simply sell in-demand films to customers at a price point they can afford.[86] In some cases, they also function as important repositories of culture, where moments of film history are captured, retained, and distributed to a local and global audience. He tells the story of Juan, who runs a stall in Tepito that specializes in "old Mexican movies" sourced from official DVDs, "8mm and 16mm reels," and "VHS and Beta recordings of TV broadcasts."[87] For Juan, this piracy functions "as a social service as well as a business, one which keeps films in circulation and keeps Tepiteño culture vibrant."[88] More than just selling Mexican films cheaply to locals, it gives life to orphan works that have been abandoned by studios that have closed down but continue to retain the film's copyright, leaving the work with no authorized method of distribution. In addition to all this, Juan sources and copies obscure films—which can sell for as much as

10,000 pesos (US$524–526 at the time of writing)—for locals and overseas collectors.[89] He has even performed similar services for the Imcine, Mexico's state-run film institute.[90] Lobato's study paints a compelling picture of Juan's piracy as a generative act, one that responds to market demand and also sustains local and national film cultures.

Conclusion

This chapter has referred to notions of use and authorship in a relatively liberal fashion. Unlike other chapters in this book, which have at least addressed case law or legislation, here we needed to move beyond the institutional framework of law to accurately examine the pirate. This departure revealed how the pirate is structured in relation to existing sites of power. The chapter also located generative aspects of piracy and discussed how infringement can sustain industries, support wider cultural and educational development, and lead to further innovation.

The examples discussed offer a number of lessons for copyright law, the legal framework that has largely been absent from this chapter. The various cultural discourses that exist around piracy underline the limitations of legal institutions, which often speak about piracy in relation to infringement. In contrast, a cultural analysis presents us with piracy that can function (and occasionally look) like authorship and also provides evidence as to how individuals become interpellated as pirates, with subjectification largely tied in to broader networks of capital and power. Therefore, it is worth paying attention to new discourses that attempt to recuperate acts of copying, such as sharing, and new "spreadable" practices that challenge or call into question how law seeks to structure modes of production, distribution, and consumption. The SOPA and PIPA protests were also instructive, as they responded to the enforcement of antipiracy laws by articulating concerns about free speech and censorship. In these cases, grounded in the context of affluent digital and social media use, we saw a strong pushback against the "piracy" label and a broad attempt to trouble the clear distinctions between authorship, use, and piracy that law seeks to maintain.

The discussion around the generative potential of piracy also presents a challenge to copyright law, namely around the philosophy that undergirds the law itself. Although copyright law can be justified by various philosophical positions,[91] one of the most commonly invoked is the utilitarian approach, which offers authors a limited monopoly as it is the best way of encouraging creativity and innovation. Yet this chapter presented a series of examples of piracy or copying that supported innovation and creativity either directly or indirectly. Therefore, we can envisage a relational pirate that not only retains some of the capacities that are presumed to reside solely in the author, but also directly challenges the structural separation of these subjects.[92] Bringing a relational approach to bear on piracy has uncovered a series of interstitial connections between subjects, as well as interesting moments of divergence between the practices of businesses, individuals, and formal law, offering an alternative narrative of piracy.

6   Producing the Pirate: The Courtroom and Cultural Power

Since the turn of the millennium a number of high-profile cases concerning online copyright infringement have been tried in different jurisdictions. These cases have been instigated by members of the film, television, and music industries in an attempt to stem the increase in online piracy taking place on and through peer-to-peer (P2P) networks. Judges have made findings with respect to the legality of these networks and have also decided on the extent to which Internet service providers (ISPs) have a role to play in stopping copyright infringement. In the examples that follow, we will see how the bulk of these decisions draw on and subsequently formalize dominant cultural narratives that have already been identified around the pirate subject. The user and pirate gradually become intertwined as legal institutions become increasingly unwilling to draw firm boundaries between these two subjects.

In the previous chapter, I argued that we needed to move beyond law to understand the pirate, and to set out a broad conceptual rethinking of the subject that had little to do with formal legal doctrine. Although legal doctrine does not address the pirate and treats infringement relatively strictly, legal institutions are not hermetically sealed off from culture. The cultural discourses around piracy that were examined in the previous chapter operate beyond law but are also in a dialogue with law, and legal institutions hear this discussion and contribute to it by producing legal interpretations of cultural life.[1]

In this chapter, we see examples of the judiciary recognizing the cultural contribution of law and making an effort to protect it. This becomes particularly apparent in cases featuring US-based companies in non-US jurisdictions. Courts want to maintain their unique ability to "produce" pirates and so are particularly unwilling to authorize extrajudicial enforcement strategies pursued by content holders. Therefore, although we can clearly see a relational pirate emerge in the last few years, it is important to note that this is due to highly strategic reasons that are related to legal institutions' attempts to retain cultural control. This chapter also challenges the scope of their ability to control culture, suggesting that much of the development around subjectivity is related to the law's struggle with unforeseen technological changes, rather than a meaningful effort to engage with the subjectivity of individuals actually affected by copyright law.

A final point to note is that this chapter has a particularly antipodean flavor as many of the cases discussed took place in Australia. This is because major content industries based in the United States have used Australia as a legal laboratory for a number of years, instigating cases that, if successful, would allow for stricter (or more effective) methods of copyright enforcement. These could then potentially be rolled out to other countries. The Australian Screen Association (ASA, though previously named the Australian Federation Against Copyright Theft, or AFACT), the local arm of the Motion Picture Association of America (MPAA), coordinated many of these cases locally. Therefore, this chapter also offers some insight into the global politics of copyright and the practice of jurisdiction-shopping, in which comparatively remote countries with assumedly more flexible law or less-developed precedent become the staging ground for an international battle over copyright enforcement. With all of this in mind, the chapter proceeds as follows. We begin by examining the history of these copyright battles, focusing on two of the most prominent cases regarding P2P technologies: *A&M Records, Inc. v. Napster, Inc.*[2] in the US (referred to in this chapter as *Napster*) and *Universal Music Australia Pty Ltd. v. Sharman Networks Ltd.*[3] (*Sharman*) in Australia. We then consider two of the more notable online piracy cases of recent years to explore how a particular jurisdiction has changed its interpretation of the pirate over the past decade or so: *Roadshow Films v. iiNet Ltd.*[4] (*iiNet*) and *Dallas Buyers Club v. iiNet Ltd.*[5] (*Dallas Buyers Club*).

A&M Records, Inc. v. Napster, Inc.

A&M Records, Inc. v. Napster, Inc. was the first time that the relatively new practice of copyright infringement through peer-to-peer networks was addressed in a courtroom setting.[6] Napster was a P2P network developed in 1999 by Shawn Fanning, a nineteen-year-old US college student. The free service allowed individuals to share MP3 files from selected folders, search other people's shared folders, and download music files. Napster continually scanned users' folders in order to retain a dynamically updated central index that users could search, and when someone clicked on a file (hoping to download it), Napster made sure the user hosting the file was online and able to send it across.[7] Then,

Napster would communicate the IP address and other relevant details of the host user to the requesting user. At that point, Napster's role in the transaction would be complete, and the actual transfer would take place directly over the internet between the hosting and requesting users.[8]

The ability to quickly search a repository of music files (both popular and obscure) and download songs for free proved to be incredibly popular, and only a year later Napster was "at its peak with twenty million users sharing six hundred thousand MP3 files."[9]

It was only a matter of time before Napster was sued. The service was publicly criticized by a number of recording artists, and record companies felt that they could not turn a blind eye to the significant amount of infringement occurring online.[10] So, in December 1999, the Recording Industry Association of America along with a number of record companies filed suit against Napster, Inc., accusing the company of engaging in "contributory and vicarious copyright" infringement.[11] The case was first heard at the US District Court for the Northern District of California, where Federal District Judge Patel granted a preliminary injunction to the plaintiffs, restraining Napster from "engaging in, or facilitating others in copying, downloading, uploading, transmitting, or distributing plaintiffs' copyrighted musical compositions and sound recordings."[12]

This was an unprecedented decision because Napster was effectively a neutral service.[13] Although it was used predominantly for copyright infringement, the peer-to-peer function at the heart of Napster could just as easily be used for non-infringing purposes. This is an important point as US copyright law had effectively refused to limit innovative technologies that could be used in substantially non-infringing ways since the *Betamax* case in 1984 (previously mentioned in chapter 2).[14] Universal Studios and Walt Disney had sued Sony, claiming that the record button on Betamax VCRs would lead to a spate of copyright infringement in homes across America. Famously, MPAA head Jack Valenti went so far as to claim that "the VCR is to the American film producer and the American public as the Boston strangler is to the woman home alone."[15] The Supreme Court disagreed, stating that "[T]he Betamax is ... capable of substantial non-infringing uses [and that] Sony's sale of such equipment to the general public" would "not constitute contributory infringement of [the] respondent's copyrights."[16] Considering this legal precedent, Napster founder Shawn Fanning, CEO Hank Barry, and their legal team sensed that they had been treated unfairly in the district court, so they appealed the decision.

The Ninth Circuit Court of Appeals reviewed the case in a narrow fashion. Rather than engaging with the wider cultural politics of copyright that emerged from the case (as we will soon), the court "reviewed legal principle within the [doctrinal and conceptual] framework" outlined by Judge Patel in the original case;[17] the court of appeals was essentially checking that precedent supported the original judgment. The court found that Napster's operations were not covered under fair use and, moreover, the service not only failed to police online piracy but knowingly encouraged and assisted "the infringement of plaintiffs' copyrights."[18] This meant that Napster was engaging in contributory and vicarious copyright infringement. The only issue that the court of appeals had with the judgment was the scope of the preliminary injunction. In the end, Napster was not forced to shut down, but because the injunction required Napster to filter copyrighted content from its system, the service quickly started to lose popularity.[19] The company eventually went bankrupt and attempted to relaunch as a legal online music service.[20]

Napster stands as an important intervention in the socio-legal construction of the pirate subject because it was the first time that copyright law had to come to grips with P2P networks. In her judgment, Judge Patel noted that the decision would directly address "the boundary between sharing and theft, personal use and the unauthorized worldwide distribution of copyrighted music and sound recordings."[21] It would also by necessity significantly develop legal subjectivity in copyright, as the way that legal institutions interpreted these boundaries would directly affect how legal subjects were positioned in relation to this emerging digital media ecology. As we will see, in addition to producing a new conceptualization of the pirate, which influenced understandings of the author and the user, the case also showed the legal system attempting to intervene in broader cultures of music consumption. Due to the narrow strictures of the appeal hearing, our analysis will focus on the district court hearing.

Judge Patel found that individuals were engaging in copyright infringement on Napster, and that "facilitating the unauthorized exchange of copyrighted material was a central part of Napster Inc.'s business strategy."[22] In short, the court decided that Napster was essentially a pirate network. This conclusion was established in the following ways. Judge Patel first tackled the issue of whether people on Napster were infringing copyright or simply sharing music through the affordances of the Internet, which was protected under fair use doctrine. She noted that "uploading and downloading MP3 files [was] not paradigmatic commercial activity," but the "global scale of Napster usage and the fact that users avoid paying for songs that otherwise would not be free" meant that use of the program could not fall under personal use.[23] It was also noted that based on the evidence, "virtually all Napster users download or upload copyrighted files."[24] Judge Patel also found that Napster was liable for the infringing acts of its users. This was based on emails between employees of Napster, which revealed that they knew infringing music was being transferred over their network.[25] Finally, Patel noted that Napster had the "ability to supervise the infringing activity and ... a direct financial interest" in it,[26] which supported a charge of vicarious copyright infringement (citing the precedent set in *Fonovisa Inc. v. Cherry Auction Inc.*[27] and *Gershwin Publishing Corp. v. Columbia Artists Management*[28]).

In these findings we see the court gradually producing moral judgments about the people who worked at Napster, the people who used it, and the technology that supported the entire enterprise. For example, although it was recognized that P2P technology had non-infringing capabilities, the court viewed the network architecture as largely benefiting consumption that could be classed as piracy. This distinction makes little sense legally because when it comes to fair use, if one follows the *Betamax* decision, not much separates P2P technologies from earlier analog technologies in the abstract.[29] Still, Judge Patel contended that the type of copying that occurred through Napster was of a different order than analog forms that fell under fair use provisions. The potential of a reproducible file that could be circulated en masse across the world was emphatically different from videotaping a television program or sharing a record because it stood as a severe economic threat to content industries.

The actions of users were tackled in a similar fashion. As noted earlier, Judge Patel remarked that the court had to decide whether users were simply "sharing" content or engaging in "theft."[30] She ultimately found that the size and scale of Napster meant that it was impossible to argue that individuals were using the network for personal use; they were instead pirates. The conceptualization of piracy that the judgment put forward was positioned as something occurring on a mass scale, through anonymous parties, for an individual's own economic benefit, and entirely devoid of intimacy. It therefore stood in stark contrast to "sharing," which implied social interaction and friendship. The possibility that friends were sharing music using the affordances of P2P technology was not considered in detail. Nor was there much reflection on whether these activities echoed earlier liminal practices around media that were either defended as valid private uses of copyrighted material or tacitly accepted as non-enforceable acts of private infringement.[31] Kylie Pappalardo notes that this decision signaled the beginning of an increasingly common judicial tendency of courts failing to engage in detailed analyses of "users' actual downloading practices" or their "motivations for downloading," preferring to simply tar the bulk of activities on these networks as piracy instead.[32]

The Napster employees were then corralled into this moral analysis, as they had active knowledge of the activities occurring on their network. Much was made of a document authored by cofounder Sean Parker that stated Napster could not know the "real names and IP addresses" of people sharing music because they were "exchanging pirated music."[33] This was viewed as damning evidence that the company's business strategy was predominantly built around widespread piracy. The court used the document to make an important distinction. Napster was not ignoring the fact that piracy was occurring on its network but establishing and actively encouraging new forms of illicit music consumption in order to build a user base and ultimately make money. It was in effect a pirate company using a pirate network to serve pirates (see figure 6.1).

This damning interpretation was further bolstered when Judge Patel reflected on the extent of Napster's impact on the market for musical content. She found that "Napster use [was] likely to reduce CD purchases by college students, whom defendant [sic] admits constitute a key demographic," and also noted that the network raised barriers to entry for the music industry, which had "entered the digital download market very recently, or plan to enter it in the next few months."[34] This was an interesting reflection by Judge Patel, suggesting that Napster was going to be solely responsible

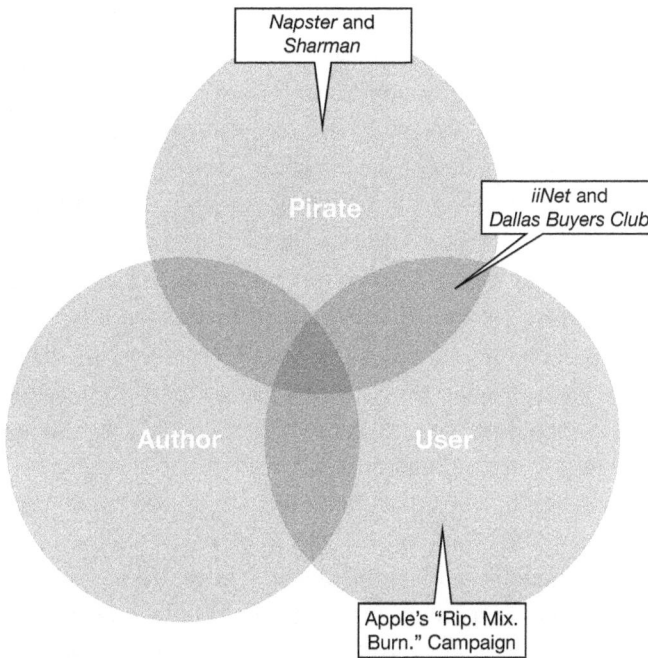

Figure 6.1
Judicial decisions and broader socio-legal discourses located infringing acts in different areas of the relational triad throughout the "copyright wars."

for a shift in the cultural consumption of music. The implication was that although people had been happy to purchase CDs prior to the launch of this service, the rise of Napster was going to fundamentally alter the market because "in choosing between the free Napster service and pay-per-download sites, consumers [were] likely to choose Napster."[35]

Judge Patel's judgment also offers an interesting interaction between law, technology, and consumption that tells us something about how the pirate subject was produced in this case. First, Judge Patel presumed that P2P technology (and more specifically Napster) was playing a central role in instigating new piratical practices of consumption. In short, peer-to-peer technology and new companies like Napster produced pirates. Second, this change was assumed to already be occurring and the natural decline of the music industry (if Napster was not curtailed) was taken as a given. Although there was evidence that copyright infringement occurred on Napster, there was no proof that the music industry was in decline, nor was it possible for legal institutions to accurately guess at the organization of the market over the years that followed. This brings us to our final point: Judge Patel felt law had a duty to intervene, to stop Napster from producing pirates, to challenge the cultural power wielded by Napster, and to respond to the existential threat it posed on behalf of the market.

This legal interpretation had some implications for subjectivity. It interpellated a swath of people who used Napster as pirates instead of users. Moreover, it presumed that unless Napster was prohibited, an increasing number of ordinary people would be interpellated in this manner once they stopped buying CDs and started using free P2P networking. This approach placed the blame for the reassignment of subjectivity on technology and companies, suggesting that the combination of these two actors encouraged people to turn toward piracy. But this was a flawed argument. In actuality, the record companies that filed the suit and the court itself immediately positioned both Napster and P2P as pirate technology and created these stark distinctions between the user and the pirate. Indeed, the music industry and the law refused to consider the possibility of relationality by not recognizing that various exceptions and limitations to copyright as well as a tacit acceptance of infringing acts had long sat in the gray areas between the user and the pirate.

Even so, the bigger question is, what did the district court seek to achieve with this legal and cultural intervention? The argument that the judgment was in consideration of the market is unconvincing, as it was unlikely that the decision was going to stop either the practice of distributing MP3 files over the Internet or the development of P2P technology. Technology innovators and entrepreneurs often flout laws in the search for greater efficiencies (and, of course, more money) and people do not read cases in detail or necessarily change their practices in response to law (as the decade of widespread online piracy that occurred following this case shows). In its judgment, the district court was contributing a performance of authority and, in doing so, recognizing the cultural power of the pirate. Judge Patel acknowledged that the power wielded by Napster was dangerous and needed to at least be addressed, even if law was always going to be on the losing side of the battle. Indeed, the decision was only the first salvo in this contest, and the cultural battle continued across the Atlantic as record companies sought to shut down one of the P2P services that appeared in Napster's wake: Kazaa.

Corporate Narratives of Digital Culture: The Kazaa
Case and the Apple Revolution

Napster was not an unqualified victory for the music industry. A number of clever entrepreneurs realized that Napster was only liable for contributory and vicarious copyright infringement because employees could see what files were being shared across their network and had the power to filter these files. This was because of the service's central index. Therefore, the solution for any enterprising P2P operator was to simply develop systems that captured little to no information and could decentralize this indexing process. The new P2P services did not retain logs of data and either "split indexing duties evenly across all peers in the network" or "limit[ed] indexing duties to certain supernodes."[36] So, as Napster reoriented its business model toward more legitimate ends, its users moved on to any of the number of peer-to-peer services that were looking to take Napster's place.

Undeterred by the limited effect of its judicial victory over Napster, the music industry filed suit against two of the more prominent new services, Grokster and Kazaa, in the US and Australia, respectively. I will focus on the Kazaa case, *Universal Music Australia Pty Ltd. v. Sharman Networks Ltd.*, because it featured a notable discussion of "pirate cultures" and the decision also went on to inform *iiNet*. The case began in a similar fashion to Napster's, with a consortium of record companies (and in this instance also individual bands and artists) filing suit against Sharman Networks, the owner of the "extremely popular" post-Napster P2P file-sharing service Kazaa, which was based out of Sydney.[37] The consortium accused the defendants of various forms of copyright infringement, but the plaintiffs' central claim was that Sharman Networks was liable for authorizing copyright infringement. In response, Sharman Networks argued that at most all it had done was distribute a piece of software that allowed copyright infringement to occur and claimed that it had no power to control the files on its network.[38]

The presiding Federal Court judge Murray Wilcox did not agree with Sharman's stance and found Sharman Networks (as well as a number of co-named defendants) guilty of authorizing copyright infringement. Regular emails and information from a marketing focus group revealed that executives knew that copyright infringement was a prominent activity on Kazaa and that people largely viewed the service "as a free music downloading search engine."[39] Moreover, despite Sharman Networks' working with a company, Altnet, to return licensed files along with unlicensed files when a person conducted a search, it was also evident that users preferred to download infringing material.[40] Further, although Judge Wilcox was not prepared to find that Kazaa had a central server like Napster, he did find that the network had the capacity to filter out particular items. Indeed, Sharman Networks had implemented a filter for adult content, but it had chosen not to introduce a similar filter for copyrighted content.[41] Subsequently, the court found that Sharman Networks had at least some "power to prevent" infringement and so "authorized" the infringing acts.

The court also found that Kazaa actively encouraged people to infringe copyright, which is of great relevance for our study of the pirate. Judge Wilcox noted that Kazaa publicly advertised a Sharman Networks campaign titled "Join the Revolution" that criticized record companies and promoted peer-to-peer file sharing on a web page. Moreover, he noted that although the campaign did not "expressly advocate the sharing of copyright files ... to a young audience ... the effect of this webpage would be to encourage visitors to think it 'cool' to defy the record companies by ignoring copyright constraints."[42] Judge Wilcox also found that Kazaa was promoted as a file-sharing facility by Sharman Networks and actively encouraged people to engage in the activity through promotions.[43] The pressing point for the judge was that "these acts took place in the context that Sharman knew the files shared by Kazaa users were largely copyright works."[44] The facts were enough for the court to find that Sharman Networks infringed copyright by "authorising Kazaa users to make copies of those sound recordings and to communicate those recordings to the public."[45] Sharman appealed to the Full Federal Court of Australia but, before the case was heard, agreed to settle out of court for a reported A\$15 million.[46]

It is particularly interesting that Sharman Networks' attempts at developing a culture of "revolutionary" file sharing was so critical to the case. Judge Wilcox was incredibly concerned with its "Join the Revolution" campaign and believed that it encouraged an already active culture of copyright infringement present on the Kazaa network. Notably, he did not consider the fact that Kazaa's manifesto outlined "methods of licensing filesharing," unlike Napster, or that there were potentially innocuous reasons for using the technology.[47] He even reflected on the personal beliefs of the people using Kazaa and stated that he did not have "any reason to believe that any significant proportion of users would care whether or not they were infringing copyright."[48] Therefore, like *Napster*, this case was not only about the technical operation of the Kazaa network but also about the fact that Kazaa appeared to produce and actively support pirates (see figure 6.1). Pappalardo notes that this judicial tendency of identifying the production of piratical subjects through technology entirely suppresses users. They are not viewed as real people but instead positioned as "caricatures of digital thieves."[49] Copyright law was once again deployed on cultural terrain, performing its authority in an attempt to stop the incessant development of pirates and reclaim its authority to produce (relatively narrowly constituted) legal subjects.

It is worth thinking more about the particular cultural interpretations that supported this judicial approach. At around the same time that *Napster* was being heard, Apple was riding high off the back of a popular advertising campaign titled "Rip. Mix. Burn." The campaign encouraged people to take advantage of the iTunes platform and the CD-RW drive built into iMacs to "rip" their music collections and make mix CDs. The campaign was clearly centered on providing people with control and creative agency over their music collections, as Apple's then-CEO Steve Jobs explained in a press release:

iMac has evolved into an entertainment center, where you can create desktop movies, manage your digital music library and burn custom music CDs ... iMovie and iTunes are so easy to learn and use, even your parents can use them without getting confused.[50]

This campaign fit squarely within the broader discourse around everyday user creativity discussed in chapter 4, as well as Apple's own corporate discourse of branding their users as "creative, non-conformist and innovative."[51] Apple was providing everyday people with the freedom to manage and reimagine their music collections, and the ability to self-identify as "creative" users.

But at the time this campaign was launched, format shifting was not legal in a number of countries. Indeed, as we saw in chapter 4, Australia, the UK, and Canada took some time to introduce an exception for format shifting into statutory law. In the US, although format shifting for personal use had always been fine, it became murkier if you shared format-shifted content (or a mix CD) with others. Therefore, the campaign received some heat from Michael Eisner, head of Disney at the time, who accused Apple of fostering piracy, and also worried the music industry, which started to use digital rights management for CDs, introducing technological copy-protection measures that made "ripping" music from CDs extremely difficult to do.[52] Yet, with that all being said, there was a distinct difference between the language used around Kazaa and Napster and that used around the Apple campaign. Although Apple customers were engaged in copying that depending on jurisdiction and the spread of the content could be considered piracy, Apple presented the acts of ripping and mixing as an evolution of traditional modes of fandom rather than a revolution in the vein of Kazaa's campaign. All the people who were ripping, mixing, and burning were gradually interpellated by culture as authorized users rather than pirates (see figure 6.1).

If we return to the *Sharman* judgment, we can see that many of the service's users would easily have fit into the "creative user" subject offered by Apple. For example, Kazaa's market research, which was presented as evidence in the case, revealed that many subscribers used "free downloading as a precursor to purchasing a CD" as a way of "sampling the full complement of a CD's songs on-line first."[53] And like users of Apple iMacs, Kazaa subscribers also engaged in ripping, mixing, and burning by downloading songs, burning mix CDs, and playing the CDs in other players and cars.[54] But whereas Apple was largely successful in its corporate framing of piracy, Kazaa's revolution failed.

This outcome was largely due to the relationship between culture and law and its impact on legal subjectivity. Apple, Napster, Kazaa, and others offered a range of new ways to consume and distribute music, allowing many individuals to infringe copyright law (often by engaging in quite similar practices). Although its campaign met with some resistance, Apple was also recognized as an innovative company on the rise, and it had mobilized an array of notable artists to support its campaign, from Barry White to Smash Mouth. In contrast, the extreme existential threat that peer-to-peer networks represented to the music industry led to a number of lawsuits and allowed a legal discourse of infringement and piracy to emerge around P2P users. The wider cultural contexts in which these revolutions took place had a real effect on how each company and its users were interpellated by law (if law engaged with these discourses at all, as it declined to do in Apple's case). Despite these divergent legal discourses, what we see in practice are in fact close links between the pirate and the user. Apple's ability to reposition copyright infringement as a creative act by iMac users and Judge Wilcox's attempt to redefine users as pirates in *Sharman* highlight how different market actors can interpellate individuals as users or pirates according to particular legal and cultural discourses.

The Fifty-First State: The United States, Australia, Free Trade, and the Copyright Experiment

Before we continue our examination of the pirate we need to briefly consider geopolitics. The next two cases we examine are also from Australia, but as flagged in this chapter's introduction, the emphasis on this jurisdiction is not due to chance or convenience. Australia has functioned as a legal petri dish for large media companies in the United States for a number of years. The MPAA has prevailed on its local arm the ASA to instigate cases in Australia on its behalf, and large multinational companies have similarly reached out to their network of Australian film and television companies and asked them to do the same. Australia is the ideal location to conduct this sort of legal activity for a number of reasons. First, the close alliance that the two countries share means that the Australian government regularly supports the strong protectionist rhetoric of the United States around IP, despite the fact that Australia is a net importer of intellectual property.[55]

Second, in recent years, content industries have viewed Australia as the most amenable jurisdiction in which to attempt to sue intermediaries. After *Napster*, a number of cases were brought against P2P software providers in the US through the secondary liability principles of US law.[56] Throughout these cases there were points of disagreement around issues of precedent, however, such as the extent of the "scope and application of the *Sony* protection," leading to "inconsistent" legal interpretation.[57] The law was also unable to grasp the difference between the physical and digital worlds, allowing new P2P networks to exploit "loopholes" in existing law.[58] The UK and Canadian jurisdictions offered no respite. Their authorization laws—broadly equivalent to secondary liability laws—were narrowly constituted and content industries would have risked a public defeat at trial if they had engaged in litigation there.[59] In contrast, authorization laws in Australia were broadly constituted in both statutory law and through jurisprudence, so content industries were more confident of not risking an embarrassing public defeat at trial if they engaged in litigation there.

The Australian and United States Free Trade Agreement (AUSFTA) stands as the final piece in this political and legal puzzle. The AUSFTA was one of the most significant copyright reform efforts ever undertaken by Australia. The agreement required that Australia adopt numerous elements of US IP law, extending its copyright term but also acceding to a range of strong enforcement measures such as the criminalization of the manufacture, use, and distribution of tools that could circumvent technological protection measures.[60] Most interestingly for our purposes, the agreement also required Australia to import a "safe harbor" regime from the United States for local ISPs. This law protected a defendant from liability for authorizing infringement if they simply provided "facilities for making, or facilitat[ed] the making of, a communication," in theory protecting an intermediary (like an ISP) from the infringing activity occurring on its network.[61]

This legal protection came at a cost. In exchange, intermediaries (or carriage service providers) would be required to adhere to a notice and takedown process. Among various other conditions, the provider would need to "implement a policy that [provided] termination, in appropriate circumstances, of the accounts of repeat infringers" and comply with any relevant industry codes that were in force.[62] By the mid-2000s content industries were slowly coming to accept that there was no easy way of stopping online piracy and that endless court cases were not going to solve the problem. Therefore, this trade agreement offered two solutions. Ideally, it would introduce a notice and takedown process along the lines of the regime in the United States' Digital Millennium Copyright Act, but it would also offer the possibility of returning to the courtroom if intermediaries did not adhere to its conditions.

But there was a major flaw in the plan. The AUSFTA did not operate as cleanly as the US notice and takedown system on which it was based. US law is clear and simply requires intermediaries to "remove" infringing material at the behest of the copyright holder or their representatives.[63] The country also has a broad definition of "intermediary" and courts have granted "safe harbor" protection to search engines, ISPs, email providers, and social media platforms, among others.[64] In contrast, Australian law requires an intermediary to act on a policy that allows repeat infringing accounts to be terminated. It is not clear from this description, however, what such a policy entails and at what point an account needs to be terminated. As we will see in the next section, this creates a problem as intermediaries and content industries have very different opinions on what constitutes an appropriate process for the termination of an account.

The concerted effort by a number of US content industries to strategically proceed with lawsuits in friendly jurisdictions is a natural extension of the broader practice of "forum shopping" or "regime shifting," in which the United States and Europe gradually moved "intellectual property lawmaking from WIPO [World Intellectual Property Organization] to GATT [General Agreement on Tariffs and Trade] to TRIPS [(The Agreement on) Trade-Related Aspects of Intellectual Property Rights]," expanding the scope of intellectual property protection in the process.[65] In a similar fashion, the following cases are not merely examples of rights holders enforcing their copyrights but rather should be read as attempts to establish new enforcement regimes through the courts and to contribute to a broader public relations exercise: showing the general public that copyright holders are serious about tackling online piracy. In short, these lawsuits are not just discrete civil actions but part of a broader geopolitics of copyright enforcement driven largely from the shores of the United States.

iiNet and *Dallas Buyers Club*: The Latest Round in the Piracy War

The *iiNet* and *Dallas Buyers Club* suits were brought in response to the fact that people were continuing to engage in significant levels of online copyright infringement. The music industry had won many of the P2P battles of the 2000s, but people continued to download music and were starting to download entire films and television series with relative impunity. This was largely thanks to the rise of BitTorrent, a novel P2P program that distributed the responsibility of uploading files to the network across several users. This drastically sped up the ability to transfer files and users could now "quickly upload and download enormous amounts of data, files that [were] hundreds or thousands of times bigger than a single MP3," building up significant personal media libraries in the process.[66] This change in piracy dragged the Western film and television industries directly into the fray and meant that all of the major content industries were collectively committed to dealing with the piracy problem as quickly as possible. Their strategy involved moving past peer-to-peer networks themselves and focusing on more stable actors that could be clearly identified and charged. *iiNet* targeted an intermediary—the Internet Service Provider—and *Dallas Buyers*

Club went one step further and targeted the infringing individual. I will briefly outline how each case proceeded and then analyze the collective implications of these decisions for the pirate subject.

In *iiNet*, a consortium of thirty-four Australian film and television companies accused an Australian ISP, iiNet, of authorizing the copyright infringement that occurred on its network. The case is important for a number of reasons. Most notably, it was "the first time that a court at the apex of a national legal system [had] considered the liability of an ISP for infringements committed by its subscribers."[67] It was also the latest battleground in the content industries' global war on piracy. Although the plaintiffs were a collection of Australian film and television companies, the Motion Picture Association of America "was the mover behind [the] case."[68] The suit was advanced by AFACT (now the ASA) in order to obscure the fact that a body representing the "vibrant ... American motion picture, home video and television industry" had pushed its local representatives to file suit.[69]

In addition to locating a friendly jurisdiction, AFACT and the MPAA picked their target carefully. A confidential diplomatic cable written by the US Ambassador to Australia at the time, Robert McCallum Jr., explained that the MPAA wanted to avoid tangling with Telstra—Australia's former telecom monopoly—whose own ISP, BigPond, had cornered half the market.[70] Telstra had the financial resources to maintain a long and potentially damaging legal battle. In comparison, as the third largest ISP in Australia, iiNet was large enough for the case to have some impact but small enough that the ISP lacked the resources or energy to defend a well-financed legal campaign headed by an assortment of national and multinational entertainment companies. Moreover, the fact that "iiNet users had a particularly high copyright violation rate, and that its management ha[d] been consistently unhelpful on copyright infringements" stood as further motivation for the MPAA to pursue legal action.[71] With all this careful preparation, AFACT and the MPAA can be forgiven for being quietly optimistic when the case began.

The case was first heard in the Federal Court of Australia in 2009, where Justice Dennis Cowdroy had to decide "whether [iiNet] had, by failing to take any steps to stop infringements, authorised the infringements of particular iiNet subscribers."[72] AFACT had sent weekly emails to Michael Malone, the managing director of iiNet from June 2008 to August 2009, alleging that particular iiNet users had infringed copyright.[73] A spreadsheet was attached to each of these emails detailing the date and time of infringement, the IP address of the customer, the particular copyrighted material downloaded, and the studio to which the copyright was attached.[74] AFACT claimed that per the terms and conditions of iiNet's customer relationship agreement and in accordance with the notice and takedown provisions introduced by the AUSFTA, iiNet should have disconnected these users rather than continuing to serve them.[75]

iiNet countered by arguing that the AFACT notices carried no legal weight and that any concerns should have been passed on to the relevant authorities.[76] It also noted that the mere provision of the Internet could not be seen as inducing or even authorizing copyright infringement, and that the infringements took place exclusively through the BitTorrent system, which was not an illegal protocol, nor under iiNet's control.[77] Furthermore, by cutting off a user's access to the Internet, iiNet would not just be cutting off the means of infringement but also access to a host of other services, such as Internet banking, news, and email, and this would represent a disproportionate punishment, especially when based on evidence that was simply presented to the ISP and not yet tested in court.[78] Justice Cowdroy found in iiNet's favor, supporting many of these arguments, and stated that he did not believe iiNet had provided the "means" of infringement.

Undeterred by this decision, the collection of production companies marshalled by AFACT appealed to the full bench of the Federal Court. Their appeal was heard by Justices Emmett, Jagot, and Nicholas in 2011 and was dismissed. The film and television companies then appealed to the High Court. The court agreed to hear the appeal and the hearings began in late November 2011; four months later on April 20, 2012, the court unanimously dismissed AFACT's appeal. The majority judgment of Chief Justice Robert French and Justices Susan Crennan and Susan Kiefel supported iiNet's submission, finding that AFACT's interpretation of authorization assumed "obligations on the part of an ISP which the Copyright Act [did] not impose."[79] The justices also found that the AFACT warning notices "did not provide iiNet with a reasonable basis for sending warning notices to individual customers containing threats to suspend or terminate those customers' accounts."[80] As a concluding point, the justices noted that "the statutory tort of authorisation of copyright infringement" was "not readily suited to enforcing the rights of copyright owners in respect of widespread infringements" and suggested that a legislative intervention was the best way of stemming online piracy.[81]

In the *Dallas Buyers Club* case, the copyright holders of the film *Dallas Buyers Club* (or DBC) made a preliminary discovery application to an Australian court. They had identified 4,726 IP addresses that were sharing the movie on BitTorrent without permission and were asking for ISPs to be legally compelled to pass on the personal details of the individuals associated with each address.[82] A group of ISPs (of which iiNet was one) challenged the application, arguing that the evidence provided was not sufficient, the claim was speculative, the court should not "order them to divulge their customers' personal and private information," and "the monetary claims which the applicants had against each infringer were so small that it was plain that no such case could or would be maintained by the applicants."[83] Finally, the defendants contended that if the court required them to divulge the information, the privacy of the account holders should be "adequately protected" and the copyright holders should not send them "speculative invoices."[84]

Speculative invoicing was one of the central reasons why these ISPs refused to release the account information. This common enforcement action involves copyright holders (or their representatives) sending letters that accuse individuals of copyright infringement. The letter usually offers the recipient the possibility of coming to a financial settlement with the copyright holder and the suggested sum is a significant amount of money, often around a few thousand dollars. Importantly, these letters suggest that if a settlement is not reached, further (and more expensive) legal action may ensue. This strategy has been criticized for essentially functioning as an income stream for copyright holders rather than a genuine attempt at deterrence, for charging sums incommensurate with the copyrighted material, and for interfering with an individual's right to due legal process.[85] The tactic was common in the US and the UK but had not yet found favor in Australia. It was suspected that *Dallas Buyers Club* was the first attempt to export that practice to Australia.

In April 2015, Justice Nye Perram heard the case at the Federal Court of Australia and found for the plaintiffs, allowing preliminary discovery to take place. He was mindful of the concerns raised by the ISPs, however, so even though he required "the information to be provided," he imposed "safeguards to ensure that the private information remain[ed] private" and "to constrain the use to which the information [could] be put."[86] The plaintiffs would only be able to use the private information to identify, sue, and negotiate "with end-users regarding their liability for infringement."[87] When the parties reconvened to hear the final order, Justice Perram stipulated a further constraint. In order to guard against speculative invoicing, he required the plaintiffs to provide "the form of any letter they intend[ed] to send to account holders" to the court before they would be allowed to contact them.[88] The first draft letter was put forward and suggested that recipients contact a representative of DBC to discuss the matter further. In a hearing in June, Justice Perram made it clear that he would not approve a letter unless he was "shown what it was that DBC was proposing to demand monetarily or ... the methodology underlying its approach to the amounts it was going to claim."[89]

DBC eventually submitted a letter that detailed the amount it would claim from each subscriber. The ISPs claimed this letter was yet another example of speculative invoicing. The parties returned to the courtroom in August of the same year to hear Justice Perram's opinion of the letters and the damages sought. Once again, Justice Perram did not allow the letters to be sent out. He noted that it was fair for DBC to make a "claim for the cost of an actual purchase of a single copy of the Film for each copy of the Film downloaded" and for damages arising from the money it cost to access the infringer's name.[90] But Justice Perram challenged DBC's attempt to claim "a one-off licence fee from each uploader on the basis that each was engaged in the widespread distribution of the film," noting that the idea that people would approach "DBC to negotiate a distribution arrangement in return for a licence fee [was] so surreal as not to be taken seriously."[91] More pressingly though, he noted that "DBC ha[d] made no submission to [him] about how these damages might be calculated or what [it would] seek."[92] In closing, Justice Perram did not only refuse to lift the stay on the private information but also required DBC to submit a A$600,000 bond, so that if it went ahead with making excessive financial demands of infringers in contempt of the court, "it [would] not be profitable for it to do so."[93]

In December 2015, DBC attempted to lift the stay one more time. The plaintiffs maintained that they should be able to charge infringers a license fee, asked to submit a reduced bond of A$60,000, and proposed to narrow the scope of their claim and target only the account holders of iiNet.[94] Once again, Justice Perram was not moved. He refused to reconsider the license fee issue and noted that DBC could not claim additional damages from subscribers on the basis of its current evidence.[95] Since DBC was still proposing to go beyond the constraints Justice Perram had set down in August, he did not address the application to lower the bond and refused to let DBC contact subscribers.[96] He ended the hearing by stating that proceedings would terminate in February 2016 unless DBC made a serious effort to explain how it was going to contact customers while staying in line with his August judgment. DBC did nothing after that point and the case was dismissed.[97]

When it comes to subjectivity, these cases function in a similar way to the earlier *Napster* and *Sharman* cases. In both *iiNet* and *Dallas Buyers Club*, legal institutions wanted to make sure they retained the sole ability to produce pirate subjects. Content industries were essentially hoping to get a legal rubber stamp for the novel enforcement measures they had devised, which would then allow them to enforce online piracy outside the courts. If the consortium of film and television companies had been successful in *iiNet*, Australian ISPs would have been compelled to cancel the accounts of subscribers that were accused of online copyright infringement. Similarly, if the copyright holders of *Dallas Buyers Club* had won their case, they (and other copyright holders) would have been able to repeatedly apply for preliminary discovery through the courts and then send invoices to suspected infringers. In both cases, what was at issue for copyright law was both legal and geopolitical. In a legal sense, the courts were reluctant to authorize enforcement mechanisms that would significantly affect individuals and devolve the evidential and decision-making processes to content industries. Tied to this was also the fact that companies and industry bodies from the United States had advanced the cases. Therefore, this production of subjectivity would not just have devolved from the courts to industry but would also have ended up moving from Australia to the United States.

In addition to protecting their own interests in the production of subjects, the courts also signaled a willingness to develop their own understanding of the pirate as a subject. Both cases considered the rights and broader actions of allegedly infringing ISP subscribers, in contrast to how law had approached users of Napster or Kazaa. In *iiNet*, the Federal and High Courts were wary of approving an enforcement mechanism that would interpellate individuals solely as pirates and disconnect their Internet use, irrespective of the fact that individuals required the Internet for a variety of everyday activities.[98] Instead, the Federal Court articulated an image of a user–pirate, noting that although AFACT offered a picture of people continually flouting copyright law, based on the evidence presented in court, piracy existed as a minor media practice that was occurring alongside a number of other mundane activities such as emailing, listening to radio online, and streaming television. In a similar fashion, Justice Perram consistently treated prospective pirates as rogue users throughout *Dallas Buyers Club*, consciously protecting them from excessive settlements and noting that "it should not be assumed that every [letter] recipient has engaged in infringement."[99] In contrast to this nuanced approach, the proposed enforcement regimes of canceling the ISP subscriptions of individuals or sending speculative invoices would routinely produce pirates with no due process and no consideration of the gray areas around this subject (see figure 6.1).

Both decisions were laudable because not only did the courts retain a careful understanding of the situated and relational nature of the pirate subject, but they also refused to intervene in wider policy processes, suggesting that many of the issues presented to the court should be considered by the legislature instead. Since we know that people do not care about the nuances of legal decisions, it is worth reflecting on what these careful judicial approaches to subjectivity mean. Throughout this chapter, we have seen that the courts are loath to allow other actors or processes to engage in legal subjectification. But as seen in the earlier example of Apple's "Rip. Mix. Burn." campaign as well as in discussions throughout the book, we also know that legal subjectivity is as much a cultural as it is a legal process. Therefore, I suggest that in these last two Australian cases, the courts were actually following culture rather than intervening and seeking (ambitiously) to lead cultural change as they did with Napster and Kazaa.

The cultural differences between these cases were stark and ultimately related to the role of the Internet in our everyday lives. During the time of *Napster*, the Internet was a relatively popular but in no way essential communication technology. Even around the time of *Sharman*, one could conceive of a moment when they would be "offline." Today, this is no longer possible. The emergence of smartphones and the mainstream adoption of social media mean that we are living media lives.[100] Therefore, it is worth considering how the pirate and copyright law more generally operate (both explicitly and implicitly) with reference to this changing aspect of culture. Napster users were seen to be damaging an as-yet-unrealized future for online music as well as flagrantly ignoring their ability to purchase music from a still-vibrant CD market. Their consumption and subjectivity were constituted in relation to a cloistered pirate network and compared to an offline or analog existence. We were all online by the time of *iiNet* and *Dallas Buyers Club*, so at no point during these judgments were people's infringing activities wholly separated from the daily consumption of varied and diverse forms of media. In a strange way then, with regard to the notion of subjectivity, the decisions were made before these final two cases were even heard. At the time of judgment, Australians consumers were both pirates and users, and the courts could not ignore this cultural context as they formulated their decisions.

From Bazaars and Bedrooms to Subscribing and Streaming: The End of the Pirate?

As this chapter ends and we move toward the conclusion, it is worth reflecting on some recent changes that raise serious questions about the extent to which the pirate will be maintained as a strong and viable subject in the future. The subject came into its own when the Internet functioned as a relatively disruptive force and individuals and new companies took it upon themselves to make use of a variety of new and often unauthorized distribution streams. Over the last few years, however, the music, film, and television industries have arguably been moving back to a period of stability. In many nations, services have launched that allow people to stream music, film, and television for a relatively low price. It is unclear what macro effect this has had on infringement but there has been a growth in the purchase and use of legal media consumption platforms and a perceived decline in piracy.[101] Furthermore, countries like the UK and Australia are now asking ISPs to block major pirate sites (such as The Pirate Bay), making it increasingly difficult (but not impossible) for casual pirates to infringe copyright.

Indeed, many artists have moved on from the fight against copyright infringement, with Radiohead and, more recently, comedian Louis C.K. releasing material through a "pay what you want" system in an attempt to remove the assumed major draw of piracy: the fact that you can access popular media for free.[102] The music industry is also starting to embrace streaming services like Pandora and Spotify and to monetize popular music and music videos on YouTube.[103] Existing revenue streams have been enhanced, with musicians making significantly more money from live shows and growing festival cultures encouraging fans to purchase an experience rather than a cultural artifact like a digital file.[104] In a similar fashion, streaming platforms such as Netflix and Hulu offer film and television companies new business models through which to generate income.

In this context, media companies have become slightly more hopeful and consumers less antagonistic. There has been some suggestion that we have started to move away from piracy and that soon we will become law-abiding consumers again. But has consumption really changed across the past decade and a half or has industry just given up? Brave experiments with streaming platforms have provided some life to the major industrial structures of the content industries (such as record companies), but recording artists do not earn much from them.[105] Moreover, the phenomenon of streaming has essentially mimicked the experience of piracy, providing audiences with a rich buffet of content with the very minor proviso that they will have to pay a small price to access it. In short, this period of stabilization has essentially seen the content industries come relatively close to matching piracy's original terms: lots of content for a negligible price. Finally, a new battle has emerged around the practice of geo-blocking, in which these new platforms geographically restrict content in line with the licensing agreements they have made with content providers. In response, many people have purchased consumer-oriented virtual private networks that allow them to surf the Internet as though they are in another location and flout these barriers with relative ease.[106] Indeed, the practice of geo-evasion may soon lead to the decline of the long-held practice of geographical windowing of content.

At first glance, it appears that pirates may have won the war after all. Despite the numerous legal challenges detailed throughout this chapter, there was very little demanded from pirates in the end. There was no real change in their scope of consumption—instead, the necessity of a smorgasbord of content was taken as a given—and it only took a move from The Pirate Bay to Netflix and the payment of a few dollars a month for these pirates to become users. Considering the similarity, perhaps pirates were just under-served audiences all along, essentially users re-interpellated by industry and occasionally by law while they waited for the music, film, and television industries to provide the type of services promised way back in the *Napster* case. This relational aspect of piracy appeared in *iiNet* and *Dallas Buyers Club*, as copyright law started to reassess the way these actors had been considered and to account for the wider cultural changes that had occurred around them. Legal institutions recognized that pirates were also users and that their demands needed to be accounted for in any judgment made, a notable discursive shift from the way individuals using Napster had been constituted.

Still, it is worth recognizing that the pirate remains an active actor in the international arena, with a range of multilateral and bilateral trade agreements seeking to interpellate a criminal pirate subject. For example, the multinational Anti-Counterfeiting Trade Agreement (ACTA) features the emotive phrase "pirated copyrighted goods"[107] and extends criminal liability to piracy on a commercial scale even if this sort of copyright infringement involves "commercial activities for … indirect economic or commercial advantage,"[108] which, as Michael Carrier notes, is "not defined in ACTA."[109] Considering the value ascribed to intellectual property globally as an economic unit and the preponderance of treaties focused on protecting IP, the pirate should retain its presence as a subject in international law even though it has become somewhat dormant in domestic jurisdictions. Moreover, the notion that the problem of access has been entirely solved is largely a US-centric view. Despite the emergence of various legal platforms, the majority of audiences from other countries still need to engage in some unauthorized activity in order to access particular forms of content, which suggests that piracy will continue for some time yet.[110] In addition to this, the prospective domestication of 3D-printing technologies and the emergence of live streaming as a method of content distribution, among other technological developments, suggest that debates around piracy and intellectual property will be ongoing.[111] Practices are gradually trending toward a more authorized path but it is highly doubtful we have seen the last of the pirate.

Conclusion

Authors, Users, and Pirates has conducted a relational analysis of copyright law and subjectivity. The book examined the three central subjects of copyright simultaneously (the author, the user, and the pirate), and outlined how they were established, explored the processes of subjectification that turn individuals into subjects of copyright law, and argued that these subjects were interrelated. The analysis was articulated through a relational framework that mapped the relationships between these subjects and followed how these subjects changed in response to legal reforms, judicial decisions, and broader socio-legal discourses. The framework also accounted for the subjects that operate in the spaces between these formal subjects, which are often produced at moments of transition.

Scholars of intellectual property know that copyright is relational. The field recognizes that individuals can simultaneously be authors and users or users and pirates (and so on). Yet the broader notion of relationality had not been substantially theorized or meaningfully recognized until recently when Carys Craig pointed out that copyright is structured around a "pervasive individualism" that inhibits communication and cultural development.[1] *Author, Users, and Pirates* has used this theoretical intervention to critically examine subjectivity and, in doing so, offers a number of useful contributions to the ongoing scholarly debate on the role and function of copyright law, as well as further developing this relational turn in copyright scholarship.

Authors, Users, and Pirates has revealed that these subjects have regularly intersected with one another as different socio-legal cultures have responded to technological change, legal decisions, and the everyday practices of individuals. For example, in chapter 4 we saw how a growing cultural conversation around the potential of amateur media creation led to a reconsideration of the user in law. Various jurisdictions had linked the user closely to the pirate through a relatively author-dominated interpretation of copyright law. This cultural conversation encouraged Australia and Canada to introduce a range of laws that emphasized the authorial capacity of the user and supported an interpretation of a user that was more than just a consumer. Chapter 6 provided another example of the socio-legal construction of relational subjectivity, with courts gradually broadening their interpretation of the pirate throughout the ongoing "war on piracy,"[2] moving from a vision of pirates as unrepentant infringers to a more nuanced ideal of a user–pirate that engaged in a range of activities, only one of which was copyright infringement. These changes highlight the critical role that socio-legal discourses play in the constitution of subjectivity and also allow us to identify subtle moments of change when agencies usually located with one subject are granted to another.

This study also raises important questions about how copyright functions in creative practice, and its usefulness to the creative community more generally. An exploration of the lines that are drawn around authorship, use, and infringement highlighted a number of gray areas in discussions around the idea/expression dichotomy in chapter 1 and substantial similarity standards in chapter 3. These analyses underlined the essentially relational nature of creation and noted that foundational conceptual tools of legal analysis were not always attuned to these processes. Considering that one ostensible goal of copyright law is to encourage innovation and the future production of creative works, examining how law and culture limit relational creativity in particular cases becomes a necessary task. As was noted in the *Larrikin Music* decision (discussed in chapter 3), copyright law often places legal institutions in situations where they are unable to account for the kind of relational creativity on which these public policy goals rest.

The occasional inability of copyright law to make these determinations in creativity also raises questions about its ongoing relevance. We saw examples of contemporary art industries in Commonwealth countries working largely off contractual agreements, and the production of screen content similarly entailed the assistance of intermediaries in order to fairly operationalize the kind of relational creativity these industries require (for example, collecting societies that represent particular creative sectors with regard to licensing and then collect and distribute royalties among members). Moreover, discussions in chapter 5 also highlighted that the absence of copyright does not necessarily mean the absence of regulation or innovation. Chefs, fashion designers, and social media platforms and audiences viewed copying as an economic driver rather than a cost. This suggests that greater nuance needs to be introduced in copyright policy-making. Despite the range of relationally attuned modes of regulation found in the creative industries, in the policy arena the rights granted by copyright are regularly positioned like an individual property right.[3] The everyday practices of particular creative industries suggest that the ongoing relevance and value of copyright as a useful legal framework is not assured. Subsequently, optimizing copyright to better account for relationality not only offers a clearer account of creativity (as previously noted) but also may sustain copyright's efficacy in the long term.

That being said, the book has also acknowledged that copyright law currently has the capacity to recognize and even enable relational interactions. Placing a greater focus on relationality does not necessarily have to involve a significant change in how copyright law is currently structured, but rather in how it is interpreted. Indeed, *Authors, Users, and Pirates* has made the case that relationality occurs at a conceptual level and retains a somewhat common presence in copyright at a practical level. This is not to discount the real transformation that a truly relational approach to copyright law would engender at a foundational level, but a tendency toward relationality has been noted throughout.

There is not one simple way to further encourage this trend. What *Authors, Users, and Pirates* can offer to this broader task is a new theorization of copyright law as a network of subjects that are meaningfully connected to one another. At a scholarly level, this approach offers a broader understanding of subjectivity in copyright law. At the level of public policy, it challenges simple narratives advocated by various actors who talk about authors, users, and pirates as though these subjects are easily locatable categories. Indeed, the traditional divisions in public discourse between authors and pirates, copyright holders and the general public, and "righteous" local pirates and the unvirtuous piratical East fail to capture the intricacies of how copyright actually manifests in people's lives. A relational approach offers a new conceptual lens through which to consider the ongoing policy problem of establishing functional copyright laws for a digital age. It requires stakeholders to realize that: at times, infringement is not an outright wrong but part of many people's lived experience; the lines between infringement and use are not only hard to draw but often defined by everyday practice and field-specific approaches rather than formal doc-trine;[4] and genuine authorial contributions are regularly generated from these complex intersections between "infringement" and "use."

In addition to these contributions to copyright scholarship, *Authors, Users, and Pirates* has also discussed historical and contemporary media distribu-tion and consumption practices and examined how various media indus-tries have interacted with copyright law. This focus on media industries highlights just how critical particular socio-technical arrangements have been to the shaping of the author, user, and pirate. The book has focused particularly on how ongoing changes in media production, distribution, and consumption have embedded relationality in the contemporary media environment and explained how many of these developments have gone on to influence how particular subjects have been interpreted in specific jurisdic-tions and have even sparked legislative change. Considering the tight links between media industries, consumers, and copyright law, there is scope for future industry configurations to encourage further relational interactions in law. Additionally, this study gives scholars of media industries a sense of how major industry changes interact with legal frameworks like copyright law at a significant level of detail.

At this point, it is also worth noting what this book did not do. The anal-ysis was limited by its focus on four common law jurisdictions—Australia, Canada, the United States, and the United Kingdom—and so its results are not necessarily relevant to civil law jurisdictions. I have also conducted an interdisciplinary study and written with the goal of developing an interest-ing narrative that could be read across numerous disciplines, particularly law, media studies, and cultural studies. Therefore, although I have seriously engaged with legislation and cases, I have often not produced the type of detailed legal analyses that might be recognizable to legal scholars, who may produce further insights about particular cases. Finally, I should point out that a range of other intellectual property regimes and other legal arenas from trademark law to contract law have become increasingly relevant to copyright doctrine, and indeed in some cases have been deployed in favor of copyright as a regulatory mechanism to protect information. Though my study acknowledged these changes, it did not address them in detail for reasons of scope and expertise. These developments will no doubt carry implications for subjectivity in copyright law and for the future relevance of copyright more generally.

With the future in mind, I want to end with a few brief thoughts on future developments that may carry significant implications for the constitution of subjectivity in copyright law. One is the role of automation in the processing of copyright infringement claims and in searching for content. This has already been discussed in relation to YouTube's Content ID mechanism but in addition to this example we can add the process of notice and takedown introduced by the Digital Millennium Copyright Act, under which intermediaries like Google are asked to remove infringing content from search results. In an ideal world, these notices would be sent by copyright holders and carefully reviewed by employees at the intermediary to assess the claims to copyright protection. However, many rights holders outsource this enforcement method to agents who send computer-generated notices to large intermediaries in bulk. Intermediaries tend to interpret these claims conservatively, accede to these notices, and remove copyrighted content regularly.[5]

This process does not have user exceptions embedded in its procedural interpretation of copyright. Instead it approaches the use of copyrighted material in a binary fashion, as either authorized by the copyright holder or unauthorized. The user is effectively removed as a subject of consideration in the latter circumstance and must be reinscribed by either an intermediary or an individual challenging the automated removal. The embedding of these automated systems of enforcement in popular platforms like YouTube underlines the fact that copyright is always at risk of being reduced to an unproductive binary. The algorithmic turn that is changing how we access news and watch television also has real implications for how copyrighted materials are dealt with and how individuals are interpellated in this process. The possibilities of code come with a set of limitations, namely, the inability to effectively manage and assess the complex interactions that occur across authorship, use, and piracy and that develop subjects correlating to these agencies. It is questionable whether automation and computation can ever accurately interpret copyright law, which suggests that the issues raised earlier in discussions around digital rights management and technical protection measures are as relevant as ever.[6]

We could better capture these emergent sociotechnical developments in future research around copyright by drawing on a range of new methods and concepts. Copyright scholarship has traditionally been a textual field, dominated by legal scholars, historians, and other scholars from across the humanities who tend to look at legal texts and archives. As noted in chapter 4, however, as copyright moves toward the sociotechnical, further insights can be drawn from science and technology studies, a field that has often relied on detailed ethnographic studies to develop an understanding of how particular subjects are constituted in relation to technology. To this I would also add that actor-network theory, a framework stemming from this field that "establishes a radical symmetry in its treatment of actors of all kinds, refusing any agentic hierarchy between the human and non-human," could be useful.[7] This construct allows for complex networks and relations between human and non-human actors to be mapped with some level of critical detail. Future studies that integrate the technical automation previously discussed could deploy actor-network theory in order to further examine how law, culture, and code "jostle with one another to negotiate relationships and settle into stable networks."[8]

The call for ethnographic research above highlights the most pressing issue for copyright researchers, as identified by Ian Hargreaves in his "Digital Opportunity" report.[9] Copyright policy in common law countries is based on a limited empirical base. Economist Ruth Towse supports this claim, noting that what passes for evidence is generally industry "lobbynomics" or legal and scholarly discussions based on principles of natural justice rather than empirically grounded claims about impact or efficiency.[10] *Author, Users, and Pirates* opened with a recognition of the increased role that copyright plays in our everyday lives, and the lived experience of the law was discussed throughout the book. With this in mind, the paucity of evidence around copyright's effectiveness is not acceptable. No longer can a select group of stakeholders attempt to drive copyright reform with limited or selective evidence. Therefore, although this study was largely a conceptual intervention, it points toward a real need for qualitative and quantitative data about copyright in order to better understand who copyright law is for.

Still, such a search for evidence must not rely on claims of efficiency and economic gain. Copyright is a set of rights that has a serious impact on the production and circulation of knowledge and affects a diverse set of sectors from the creative industries to libraries and education. Moreover, its legal framework intervenes in cultural processes and contributes to the production of cultural resources. Subsequently, it cannot be analyzed purely through an economic lens. Indeed, any critical examination of copyright policy raises foundational and conceptual questions around compensation to artists, the public right to access knowledge, and the importance of cultural goods more generally. These issues are questions of cultural rather than economic value and can only be solved by critiquing the cultural and doctrinal justifications for copyright.[11] From the perspective of this book, this involves an ongoing investigation into the subjects that are used to structure copyright and a better critical understanding of how they operate and are deployed.

This investigation will not be easy. In closing this book, the major argument I want to put forward is that copyright is incredibly complicated. It is a byzantine and complex legal framework that is often unable to clearly regulate creation, distribution, consumption, and borrowing or provide a logical conceptual foundation for its central subjects. In such a context, *Author, Users, and Pirates* does not provide a set of clear and overarching findings about copyright and subjectivity. This would be too ambitious a goal. Instead, it stands as an early intervention into this space, sketching out avenues of inquiry, diagnosing the complexities of subjectivity, and offering a prospective interpretive framework. Future work is needed that explores how interactions between creative practice and commerce produce subjects and, at a broader level, particular narratives and beliefs about the role and function of copyright. The ultimate goal should be to develop an understanding of the author, user, and pirate in law that does not just speak to copyright doctrine or economic value but that can articulate a viable understanding of copyright as a right while also providing a useful philosophy of the role, function, and impact of copyright in a wider social context.

Notes

Introduction

1. Jessica Litman, "The Politics of Intellectual Property," *Cardozo Arts and Entertainment Law Journal* 27 (2006): 313–320. Also see Jessica Litman, *Digital Copyright* (New York: Prometheus Books, 2001) for further discussion of this tendency in a US context.

2. Matthew Rimmer, *Digital Copyright and the Consumer Revolution: Hands Off My iPod* (Cheltenham, UK: Edward Elgar Publishing, 2007) and William Patry, *How to Fix Copyright* (Oxford, UK: Oxford University Press, 2011) have provided detailed explanations of how these developments in media distribution and consumption have affected copyright law. For further discussion on the emergence of amateur digital media see Dan Hunter et al., eds., *Amateur Media: Social, Cultural and Legal Perspectives* (Oxford, UK: Routledge, 2013).

3. Of course, copyright law has always affected everyday media consumption, since creative works are understood as forms of private property. But this previously occurred at one remove from audiences and individuals, with its effects largely felt by producers and distributors.

4. Tarleton Gillespie, "Characterizing Copyright in the Classroom: The Cultural Work of Antipiracy Campaigns," *Communication, Culture & Critique* 2, no. 3 (2009): 274–318.

5. See Jessica Litman noting that "Meanwhile, representatives of the music, recording and film industries have sued the purveyors of peer-to-peer file sharing software, the Internet service providers who enable consumers to trade files, and more than 5000 individual consumers accused of making recorded music available to other consumers over the Internet" in "Sharing and Stealing," *Hastings Communications and Entertainment Law Journal* 27, no. 1 (2004): 3.

6. Patrick Burkart, *Pirate Politics* (Cambridge, MA: The MIT Press, 2014); Martin Fredriksson, "Copyright Culture and Pirate Politics," *Cultural Studies* 28, no. 5–6

(2014): 1022–1047; and Martin Fredriksson, "Piracy, Globalisation and the Colonisation of the Commons," *Global Media Journal: Australian Edition* 6, no. 1 (2012), available at http://www.hca.westernsydney.edu.au/sites/wp_gmjau/archive/v6_2012_1/pdf/martin_fredriksson_RA_6_1_2012.pdf.

7. See Turner Hopkins, *Report for Ofcom: The Value of User-Generated Content* (Warrington, UK: Ofcom, 2013); or OECD Directorate for Science, Technology and Industry: Committee for Information, Computer and Communications Policy, *Participative Web: User-Created Content*, OECD, April 12, 2007, available at https://www.oecd.org/sti/38393115.pdf.

8. Throughout this book I refer to what is occasionally called a "subject position" in cultural theory as a "subject," largely for the purposes of simplicity and readability. Subsequently, phrases like "these subjects" refer to the categories of the author, user, and pirate.

9. Under moral rights an author has the right to protect the integrity of their work and the right to be attributed (among other things).

10. See Section 29 of the Canadian Copyright Act 1985, Section 40–44 of the Australian Copyright Act 1968, and Sections 29–30 of the Copyright, Designs and Patents Act 1988 of the UK.

11. See Section 107 of the US Copyright Act.

12. Adrian Johns, *Piracy: The Intellectual Property Wars from Gutenberg to Gates* (Chicago: University of Chicago Press, 2009).

13. See Jane Ginsburg noting "that copyright doctrine on authorship, both here and abroad, is surprisingly sparse" in "The Concept of Authorship in Comparative Copyright Law," *DePaul Law Review* 52, no. 4 (2003): 1066.

14. See Julie Cohen, "The Place of the User in Copyright Law," *Fordham Law Review* 74 (2005): 347–374; Julie Cohen, "Creativity and Culture in Copyright Theory," *U.C. Davis Law Review* 40, no. 3 (2007): 1151–1205; Julie Cohen, *Configuring the Networked Self: Law, Code, and the Play of Everyday Practice* (New Haven, CT: Yale University Press, 2012); Joseph Liu, "Copyright Law's Theory of the Consumer," *Boston College Law Review* 44, no. 2 (2003): 397–431; and Kylie Pappalardo, "A Tort Law Framework for Copyright Authorisation" (PhD thesis, Australian Catholic University, 2015).

15. Johns, *Piracy*.

16. See Abraham Drassinower, "From Distribution to Dialogue: Remarks on the Concept of Balance in Copyright Law," *Journal of Corporation Law* 34, no. 4 (2009): 991–1007; Liu, "Copyright Law's Theory of the Consumer"; and Lyman Ray Patterson and Stanley W. Lindberg, *The Nature of Copyright: A Law of Users' Rights* (Athens: University of Georgia Press, 1991).

17. See Carys Craig, *Copyright, Communication and Culture: Towards a Relational Theory of Copyright Law* (Cheltenham, UK: Edward Elgar Publishing, 2011); Jennifer Nedelsky, *Law's Relations: A Relational Theory of Self, Autonomy and Law* (Oxford, UK: Oxford University Press, 2011); and Robert Leckey, *Contextual Subjects: Family, State and Relational Theory* (Toronto: University of Toronto Press, 2008).

18. See Lisa Blackman et al., "Creating Subjectivities." *Subjectivity* 22, no. 1 (2008): 1–27; and Donald Hall, *Subjectivity* (London: Routledge, 2004).

19. See for example, Cohen, "The Place of the User in Copyright Law," and Ginsburg, "The Concept of Authorship in Comparative Copyright Law."

20. Martha Woodmansee, "The Genius and the Copyright: Economic and Legal Conditions of the Emergence of the 'Author,'" *Eighteenth-Century Studies* 17, no. 4 (1984): 425–448; Martha Woodmansee, *The Author, Art, and the Market* (New York: Columbia University Press, 1994); and Martha Woodmansee and Peter Jaszi, *The Construction of Authorship: Textual Appropriation in Law and Literature* (Durham, NC: Duke University Press, 1994).

21. Peter Jaszi, "Toward a Theory of Copyright: The Metamorphoses of 'Authorship,'" *Duke Law Journal* 40, no. 2 (1991): 455–502.

22. Mark Rose, *Authors and Owners: The Invention of Copyright* (Cambridge, MA: Harvard University Press, 1993).

23. Michel Foucault, "What Is an Author" in *Aesthetics, Methods and Epistemology*, ed. James D. Faubion, 205–222 (New York: The New Press, 1998).

24. Roland Barthes, *Image-Music-Text* (London, Fontanta: 1977).

25. Alan Durham, "The Random Muse: Authorship and Indeterminacy," *William and Mary Law Review* 44, no. 2 (2002): 569–642.

26. Oren Bracha, "The Ideology of Authorship Revisited: Authors, Markets, and Liberal Values in Early American Copyright," *Yale Law Journal* 118, no. 2 (2008): 186–271.

27. Ibid.

28. Kathy Bowrey, "Law, Aesthetics and Copyright Historiography: A Critical Reading of the Genealogies of Martha Woodmansee and Mark Rose," in *Research Handbook on the History of Copyright Law*, ed. Isabella Alexander and Tomás Gómez-Arostegui (Cheltenham, UK: Edward Elgar Publishing, 2016), 50.

29. Ginsburg, "The Concept of Authorship in Comparative Copyright Law."

30. Ibid. See also Bracha, "The Ideology of Authorship Revisited"; Rose, *Authors and Owners*; Woodmansee and Jaszi, *The Construction of Authorship*; and Christopher Buccafusco, "A Theory of Copyright Authorship," *Virginia Law Review* 102, no. 5 (2016): 1229–1296.

31. Patterson and Lindberg, *The Nature of Copyright*.

32. Ibid., 4.

33. Liu, "Copyright Law's Theory of the Consumer."

34. Johanna Gibson, *Creating Selves: Intellectual Property and the Narration of Culture* (Aldershot, UK: Ashgate Publishing, 2006).

35. Alina Ng, "When Users Are Authors: Authorship in the Age of Digital Media," *Vanderbilt Journal of Entertainment and Technology Law* 12, no. 4 (2010): 853–888.

36. Burkart, *Pirate Politics*; Fredriksson, "Copyright Culture and Pirate Politics"; and Fredriksson, "Piracy, Globalisation and the Colonisation of the Commons." Also see Kavita Philip, "What Is a Technological Author? The Pirate Function and Intellectual Property," *Postcolonial Studies* 8, no. 2 (2005): 199–218; and Lars Eckstein and Anja Schwarz, *Postcolonial Piracy: Media Distribution and Cultural Production in the Global South* (London and New York: Bloomsbury Publishing, 2014).

37. Jonas Andersson Schwarz and Patrick Burkart, "Piracy and Social Change— Revisiting Piracy Cultures," *International Journal of Communication* 9 (2015): 794.

38. Lawrence Liang, "Cinematic Citizenship and the Illegal City," *Inter-Asia Cultural Studies* 3, no. 6 (2005): 366–385; Brian Larkin, "Degraded Images, Distorted Sounds: Nigerian Video and the Infrastructure of Piracy," *Public Culture* 16, no. 2 (2004): 289–314; Brian Larkin, *Signal and Noise: Media, Infrastructure, and Urban Culture in Nigeria* (Durham, NC: Duke University Press, 2008); Ravi Sundaram, *Pirate Modernity: Delhi's Media Urbanism* (New York: Routledge, 2010); Ramon Lobato, *Shadow Economies of Cinema: Mapping Informal Film Distribution* (London: Palgrave Macmillan, 2012); and Kavita Philip, "What Is a Technological Author?"

39. Rebecca Tushnet, "Copy This Essay: How Fair Use Doctrine Harms Free Speech and How Copying Serves It," *Yale Law Journal* 114 (2004): 535–590.

40. Ibid.

41. For a discussion of this narrative see Philip, "What Is a Technological Author?"

42. Cohen, "The Place of the User in Copyright Law"; Cohen, "Creativity and Culture in Copyright Theory"; and Cohen, *Configuring the Networked Self*.

43. Cohen, "The Place of the User in Copyright Law," 349.

44. Cohen, "The Place of the User in Copyright Law"; and Cohen, "Creativity and Culture in Copyright Theory."

45. Tushnet, "Copy This Essay."

46. Patterson and Lindberg, *The Nature of Copyright*.

47. Gibson, *Creating Selves*.

48. Ng, "When Users Are Authors."

49. Philip, "What Is a Technological Author?"

50. Louis Althusser, *Lenin and Philosophy and Other Essays* (New York: Monthly Review Press, 1971).

51. Rosemary Coombe, "Room for Manoeuver: Toward a Theory of Practice in Critical Legal Studies," *Law & Social Inquiry* 14, no. 1 (1989): 69–121.

52. Thanks to Tim Laurie for assistance in clarifying these thoughts.

53. Kathy Bowrey, "Who's Writing Copyright's History?," *European Intellectual Property Review* 18, no. 6 (1996): 325.

54. Ibid.

55. Rosemary Coombe, *The Cultural Life of Intellectual Properties: Authorship, Appropriation and the Law* (Durham, NC: Duke University Press, 1998), 9.

56. Coombe, "Room for Manoeuver," 117.

57. Paul Smith, *Discerning the Subject* (Minneapolis: University of Minnesota Press, 1988), 39.

58. Ibid., 38.

59. Jennifer Nedelsky, "Reconceiving Rights as Relationship," *Review of Constitutional Studies/Revue d'etudes constitutionnelles* 1, no. 1 (1993): 7.

60. Ibid., 8.

61. Ibid., 8.

62. Ibid., 9.

63. Ibid., 9.

64. See Nedelsky, *Law's Relations*; Leckey, *Contextual Subjects*; and Jennifer Nedelsky, "Citizenship and Relational Feminism," in *Canadian Political Philosophy: Contemporary Reflections*, ed. Ronald Beiner and Wayne Norman (New York: Oxford University Press, 2001), 131–146.

65. See Carys Craig, "Reconstructing the Author-Self: Some Feminist Lessons for Copyright Law," *Journal of Gender, Social Policy & the Law* 15, no. 2 (2007): 207–268; and Craig, *Copyright, Communication and Culture*.

66. See Leckey, *Contextual Subjects*.

67. See Craig, "Reconstructing the Author-Self"; Craig, *Copyright, Communication and Culture*; also see Xiao Xiang Shi, "Towards a Relational Theory of Copyright Law: Reconfiguring Author's Economic Rights to Facilitate Knowledge Growth in Networked Information Societies" (PhD thesis, Queensland University of Technology, 2010).

68. Craig, "Reconstructing the Author-Self," 232.

69. Ibid., 237.

70. Craig, *Copyright, Communication and Culture*, 50.

71. Ibid., 51.

72. Ibid., 52.

73. Ibid., 51–52.

74. Ibid., 155.

75. See Shi, "Towards a Relational Theory of Copyright Law"; and Xiao Xiang Shi and Brian Fitzgerald, "A Relational Theory of Authorship," in *Knowledge Policy for the 21st Century: A Legal Perspective*, ed. Brian Fitzgerald and Mark Perry (Toronto: Irwin Law, 2011), available at https://www.irwinlaw.com/content_commons/knowledge _policy_for_the_21st_century.

76. Nedelsky, "Reconceiving Rights as Relationship," 2.

77. Nedelsky, *Law's Relations*, 374.

78. See Cohen, "The Place of the User in Copyright Law"; Ginsburg, "The Concept of Authorship in Comparative Copyright Law"; Liu, "Copyright Law's Theory of the Consumer"; Pappalardo, A Tort Law Framework for Copyright Authorisation; and Tushnet, "Copy This Essay."

79. Litman, *Digital Copyright*; and Blayne Haggart, *Copyfight: The Global Politics of Digital Copyright Reform* (Toronto: University of Toronto Press, 2014).

80. Bowrey, "Law, Aesthetics and Copyright Historiography"; and Ginsburg, "The Concept of Authorship in Comparative Copyright Law."

81. Craig, *Copyright, Communication and Culture*, 3.

82. Thomas Streeter, *The Net Effect: Romanticism, Capitalism and the Internet* (New York: New York University Press, 2010), 10.

83. Ibid., 12.

84. Kathy Bowrey and Jane Anderson, "The Politics of Global Information Sharing: Whose Cultural Agendas Are Being Advanced?" *Social & Legal Studies* 18, no. 4 (2009): 482.

85. Cohen, *Configuring the Networked Self*, 17.

86. Madhavi Sunder, *From Goods to a Good Life: Intellectual Property and Global Justice* (New Haven, CT: Yale University Press, 2012).

87. Anupam Chander and Madhavi Sunder, "Copyright's Cultural Turn," *Texas Law Review* 91 (2013): 1406.

Chapter 1

1. Kathy Bowrey, "On Clarifying the Role of Originality and Fair Use in 19th Century UK Jurisprudence: Appreciating 'The Humble Grey which Emerges as the Result of Long Controversy,'" in *The Common Law of Intellectual Property: Essays in Honour of Professor David Vaver*, ed. Lionel Bently, Catherine Ng, and Giuseppina D'Agostino (Oxford, UK: Hart Publishing, 2010), 45–72.

2. For leading critiques of this absence of theorization and conceptual development in copyright law, particularly around authorship, see Jane Ginsburg, "The Concept of Authorship in Comparative Copyright Law," *DePaul Law Review* 52, no. 4 (2003): 1063–1092; and Kathy Bowrey, "Law, Aesthetics and Copyright Historiography: A Critical Reading of the Genealogies of Martha Woodmansee and Mark Rose," in *Research Handbook on the History of Copyright Law*, ed. Isabella Alexander and Tomás Gómez-Arostegui (Cheltenham, UK: Edward Elgar Publishing, 2016), 27–52.

3. *An Act for the Encouragement of Learning, by Vesting the Copies of Printed Books in the Authors or Purchasers of Such Copies, During the Times Therein Mentioned* (1710) 8 Anne, c 19.

4. Lyman Ray Patterson, *Copyright in Historical Perspective* (Nashville: Vanderbilt University Press, 1968), 32.

5. Ibid., 137.

6. See Ernest Sirluck, "Areopagitica and a Forgotten Licensing Controversy," *Review of English Studies* 11, no. 43 (1960): 260–274; or Benjamin Kaplan, *An Unhurried View of Copyright* (New York: Columbia University Press, 1967).

7. Mark Rose, "Nine-Tenths of the Law: The English Copyright Debates and the Rhetoric of the Public Domain," *Law and Contemporary Problems* 66, no. 1/2 (2003): 75–87.

8. Ronan Deazley, *On the Origin of the Right to Copy: Charting the Movement of Copyright Law in Eighteenth Century Britain (1695–1775)* (Oxford, UK: Hart Publishing, 2004), 13.

9. Ibid., 26.

10. See Deazley, *On the Origin Of the Right To Copy*.

11. Ibid., 29.

12. Ibid., 33; see also John Feather, *Publishing, Piracy, and Politics: A Historical Study of Copyright in Britain* (New York: Mansell, 1994).

13. Isabella Alexander, *Copyright Law and the Public Interest in the Nineteenth Century* (Oxford, UK and Portland, OR: Hart Publishing, 2010), 22.

14. Ibid., 23.

15. This point has been made in Alexander, *Copyright Law and the Public Interest in the Nineteenth Century*; Deazley, *On the Origin of the Right to Copy*; John Feather, *A History of British Publishing* (London: Routledge, 2006); and Brad Sherman and Lionel Bently, *The Making of Modern Intellectual Property Law* (Cambridge, UK: Cambridge University Press, 1999).

16. A number of copyright histories explore these beliefs such as Deazley, *On the Origin of the Right to Copy*; Feather, *A History of British Publishing*; and Mark Rose, *Authors and Owners: The Invention of Copyright* (Cambridge, MA: Harvard University Press, 1993).

17. Isabella Alexander, "All Change for the Digital Economy: Copyright and Business Models in the Early Eighteenth Century," *Berkeley Technology Law Journal* 25, no. 3 (2010): 1359.

18. Peter Jaszi, "Toward a Theory of Copyright: The Metamorphoses of 'Authorship,'" *Duke Law Journal* 40, no. 2 (1991): 471.

19. William St. Clair, *The Reading Nation in the Romantic Period* (Cambridge, UK: Cambridge University Press, 2004).

20. Ronan Deazley, "The Statute of Anne and the Great Abridgement Swindle," *Houston Law Review* 47, no. 4 (2010): 793–818.

21. Alexander, "All Change for the Digital Economy," 1366.

22. Ibid.

23. Gyles v. Wilcox, 26 Eng. Rep. 489 (No. 130) (1740).

24. Ibid.

25. Ibid.

26. Alexander, *Copyright Law and the Public Interest*, 170.

27. Jane Ginsburg, "'Une Chose Publique'? The Authors Domain and the Public Domain in Early British, French and U.S. Copyright Law," *Cambridge Law Journal* 65, no. 3 (2006): 642.

28. Deazley, "The Statute of Anne and the Great Abridgement Swindle."

29. Daniel Defoe, *An Essay on the Regulation of the Press* (London, 1704), 20.

30. Kathy Bowrey and Natalie Fowell, "Digging Up Fragments and Building IP Franchises," *Sydney Law Review* 31, no. 2 (2009): 204.

31. Ibid.

32. Ginsburg, "'Une Chose Publique'?," 649.

33. Gyles v. Wilcox, 26 Eng. Rep. 489 (No. 130) (1740).

34. Millar v. Taylor, 4 Burr. 2303, 98 Eng. Rep. 201, 252 (K.B. 1769); Donaldson v. Beckett 4 Burr. 2408, 98 Eng. Rep. 257 (H.L. 1774).

35. This historical narrative features in the vast majority of scholarly books covering this period. The best narrative accounts of this tension between the common law right and the statutory rights ostensibly granted by Anne are Sherman and Bently, *The Making of Modern Intellectual Property Law*; Feather, *A History of British Publishing*; and Rose, *Authors and Owners* (see, for example, chapters 4 and 5).

36. For a recent summary of and intervention into this debate, see H. Tomas Gomez-Arostegui, "Copyright at Common Law in 1774," *Connecticut Law Review* 47, no. 1 (2014): 1–57.

37. Gyles v. Wilcox, 26 Eng. Rep. 489 (No. 130) (1740).

38. Ginsburg, "The Concept of Authorship in Comparative Copyright Law."

39. Bowrey and Fowell, "Digging Up Fragments and Building IP Franchises," 207.

40. Mawman v. Tegg, 2 Russell 385 (1826).

41. Ibid., at 385, 386.

42. Ibid.

43. Ibid.

44. Ibid., at 388.

45. Ibid., at 388, 389.

46. Ibid., at 397.

47. See Catherine Seville, *Literary Copyright Reform in Early Victorian England: The Framing of the 1842 Copyright Act* (Cambridge, UK: Cambridge University Press, 1999), 241.

48. Mawman v. Tegg, 2 Russell 401 (1826).

49. Ibid., at 405.

50. Kelly v. Morris, 34 LJR 423–426 (1866).

51. Ibid., at 423.

52. Ibid.

53. Ibid.

54. Ibid., at 424.

55. Ibid.

56. Alexander, *Copyright Law and the Public Interest*, 204.

57. Morris v. Ashbee, LR 7 Eq. 34 (1868).

58. Ibid., 35.

59. Ibid., 40.

60. Collis v. Cater, 78 LT 613 1898 at 615.

61. Ibid.

62. Leslie v. Young [1894] AC 335 at 340.

63. See discussion in Hasan Deveci, "Databases: Is Sui Generis a Stronger Bet Than Copyright?" *International Journal of Law and Information Technology* 12, no. 2 (2004): 185.

64. Alexander, *Copyright Law and the Public Interest*, 232.

65. See Sherman and Bently's reading of this period, *The Making of Modern Intellectual Property Law*, specifically page 135.

66. Alfred William Brian Simpson, *Legal Theory and Legal History* (London: The Hambledon Press, 1987), 279.

67. Alexander, *Copyright Law and the Public Interest*, 282.

68. See generally Bowrey, "On Clarifying the Role of Originality."

69. Ibid., 64.

70. See Jessica Litman, *Digital Copyright* (New York: Prometheus Books, 2001); Matthew Rimmer, *Digital Copyright and the Consumer Revolution: Hands Off My iPod* (Cheltenham, UK: Edward Elgar Publishing, 2007); and William Patry, *How to Fix Copyright* (Oxford, UK: Oxford University Press, 2011).

71. U.S. Copyright Act 1802 (Amendment of 1790 Act), 2 Stat. 171 (1802). Available at http://www.copyrighthistory.org/cam/tools/request/showRecord.php?id=record _us_1802.

72. U.S. Copyright Act 1831 (Amendment of 1790 Act), 4 Stat. 436 (1831). Available at http://www.copyrighthistory.org/cam/tools/request/showRecord?id=record_ us_1831.

73. Lyman Ray Patterson and Stanley W. Lindberg, *The Nature of Copyright: A Law of Users' Rights* (Athens: University of Georgia Press, 1991), 226.

74. 17 U.S.C. § 101 (Supp. 1980), as amended by The Computer Software Copyright Act of 1980, Pub. L. 96–517, 94 Stat. 3015.

75. Christopher Buccafusco, "A Theory of Copyright Authorship," *Virginia Law Review* 102, no. 5 (2016): 1242.

76. Copyright Act 1842 5 & 6 Vict. c.45, s.15.

77. Copyright Act 1911 Geo.6 5(1911) c.46. The copyright term was subsequently extended in the Copyright, Designs and Patents Act 1988.

78. Siva Vaidhyanathan, *Copyrights and Copywrongs: The Rise of Intellectual Property and How It Threatens Creativity* (New York: New York University Press, 2003), 25.

79. See Peter Drahos, "BITs and BIPs: Bilateralism in Intellectual Property," *Journal of World Intellectual Property* 4, no. 6 (2001): 791–808; or Deborah Halbert, *Resisting Intellectual Property* (London: Routledge, 2006).

80. Buccafusco, "A Theory of Copyright Authorship," 1248.

81. Ibid.

82. Ibid., 1255.

83. Pamela Samuelson makes a case for using the terms "protectable" and "unprotectable" rather than "idea/expression" as a way of offering clarity around the reasons for these boundaries. See "Why Copyright Law Excludes Systems and Processes from the Scope of Its Protection," *Texas Law Review* 85, no. 7 (2007): 1921–1977.

84. See Jaszi, "Toward A Theory Of Copyright."

85. Ibid., 485.

86. See Johanna Gibson, *Creating Selves: Intellectual Property and the Narration of Culture* (Aldershot, UK: Ashgate Publishing, 2006).

87. Bowrey, "Law, Aesthetics and Copyright Historiography."

Chapter 2

1. Julie Cohen, "The Place of the User in Copyright Law," *Fordham Law Review* 74 (2005): 347–374; Johanna Gibson, *Creating Selves: Intellectual Property and the Narration of Culture* (Aldershot, UK: Ashgate Publishing, 2006); and Joseph Liu, "Copyright Law's Theory of the Consumer," *Boston College Law Review* 44, no. 2 (2003): 397–431.

2. Imperial Copyright Act 1911, c. 46 (UK) s2(1).

3. Ibid., s2(1)(i).

4. Ibid., s2(1)(ii).

5. Ibid., s2(1)(iii).

6. Ibid., s2(1)(vi).

7. The Copyright Law Review Committee, *Report of the Committee Appointed by the Attorney General of the Commonwealth to Consider What Alterations Are Desirable in the Copyright Law of the Commonwealth*, Australia, 1958, 7.

8. Ibid., 10.

9. Copyright, Designs and Patents Act 1988, c. 48 (UK) s29–50.

10. See ibid., s31A–F and s70. Time shifting involves people taping television or radio programs at home so they can be played at a more convenient time.

11. See Copyright Act 1968 (Cth), s40–44 and Copyright Act, RSC 1985, c. C-42, s29.

12. Copyright, Designs and Patents Act 1988, c. 48 (UK) Section s18. In comparison, it has been noted that Australia may not have a clear principle of copyright exhaustion or first sale; see Jessica Stevens, "The Secondary Sale, Copyright Conundrum—Why We Need a Secondary Market for Digital Content," *Australian Intellectual Property Journal* 26, no. 4 (2016): 179–194. The growing practice of licensing rather than selling digital goods has also significantly weakened this principle; see Rachel Ann Geist, "A License to Read: The Effect of E-Books on Publishers, Libraries, and the First Sale Doctrine," *Idea* 52, no. 1 (2012): 63–100.

13. See Kembrew McLeod and Peter DiCola, *Creative License: The Law and Culture of Digital Sampling* (Durham, NC: Duke University Press, 2011).

14. Abraham Drassinower, "A Rights-Based View of the Idea/Expression Dichotomy in Copyright Law," *Canadian Journal of Law and Jurisprudence* 16, no. 1 (2003): 15.

15. Jessica Litman makes a similar argument with respect to the role of the public domain in "The Public Domain," *Emory Law Journal* 39 no. 4 (1990): 965–1023.

16. Ibid. See also James Boyle, "The Second Enclosure Movement and the Construction of the Public Domain," *Law and Contemporary Problems* 66, no. 1/2 (2003): 33–74.

17. Drassinower, "A Rights-Based View of the Idea/Expression Dichotomy"; and Litman, "The Public Domain."

18. U.S. Const. art. I, § 8, cl. 8.

19. Jane Ginsburg offers the best articulation of the tensions between these two approaches and how they are managed in US law. She also believes that too much is made of the supposed differences between these two approaches. See "Tale of Two Copyrights: Literary Property in Revolutionary France and America," *Tulane Law Review* 64, no. 5 (1990): 991–1032.

20. 9 F. Cass 342 (C.C.D. Mass. 1841) (No. 4, 901).

21. Ibid., 100.

22. Ibid., 108.

23. Ibid., 119.

24. Ibid., 107.

25. Ibid., 116.

26. Matthew Sag, "The Pre-History of Fair Use." *Brooklyn Law Review* 76, no. 4 (2011): 1373.

27. See Sag, "The Pre-History of Fair Use," or Lyman Ray Patterson, "Folsom v. Marsh and Its Legacy," *Journal of Intellectual Property Law* 5 (Spring 1998): 431–452.

28. Leon Yankwich, "What Is Fair Use?," *University of Chicago Law Review* 22, no. 1 (Autumn 1954): 213.

29. Copyright Act of 1976, 17 U.S.C § 107. For a clear explanation of this codification process, also see Pierre Leval, "Nimmer Lecture: Fair Use Rescued," *UCLA Law Review* 44, no. 5 (June 1997): 1449–1467.

30. Leval, "Nimmer Lecture," 1453.

31. Some of the more notable studies of fair use have been: William Fisher III, "Reconstructing the Fair Use Doctrine," *Harvard Law Review* 101, no. 8 (1988): 1659–1795; Wendy Gordon, "Fair Use as Market Failure: A Structural and Economic Analysis of the 'Betamax' Case and Its Predecessors," *Columbia Law Review* 82, no. 8 (1982): 1600–1657; Lucas Hilderbrand, *Inherent Vice: Bootleg Histories of Videotape and Copyright* (Durham, NC: Duke University Press, 2009); Michael J. Madison, "Rewriting Fair Use and the Future of Copyright Reform," *Cardozo Arts & Entertainment Law Journal* 23, no. 2 (2005): 391–418; and Rebecca Tushnet, "Copy This Essay: How Fair Use Doctrine Harms Free Speech and How Copying Serves It," *Yale Law Journal* 114 (2004): 535–590.

32. Patricia Aufderheide and Peter Jaszi, *Reclaiming Fair Use: How to Put Balance Back in Copyright* (Chicago: University of Chicago Press, 2011), 40.

33. Tushnet, "Copy This Essay," 545.

34. Numerous scholars have made this general argument, but interesting arguments are advanced in Tushnet, "Copy This Essay," and McLeod and DiCola, *Creative License*.

35. See Aufderheide and Jaszi, *Reclaiming Fair Use*, and Jeannine Marques, "Fair Use in the 21st Century: Bill Graham and *Blanch v. Koons*." *Berkeley Technology Law Journal* 22, no. 1 (2007): 331–354.

36. Aufderheide and Jaszi, *Reclaiming Fair Use*, 73. See also Lennon v. Premise Media Corp., 556 F. Supp. 2d 310, 328 (SDNY 2008).

37. Aufderheide and Jaszi, *Reclaiming Fair Use*, 77.

38. Ibid., 81. Also see Bill Graham Archives v. Dorling-Kindersley Ltd., 448 F.3d 605 (2d Cir. 2006).

39. Aufderheide and Jaszi, *Reclaiming Fair Use*, 73.

40. Marques, "Fair Use in the 21st Century," 333.

41. Lawrence Lessig, *Free Culture: How Big Media Uses Technology and the Law to Lock Down Culture and Control Creativity* (New York: Penguin Press, 2004), 107.

42. Amy Adler, "Fair Use and the Future of Art," *New York University Law Review* 91, no. 3 (2016): 559–626; and Rebecca Tushnet, "Worth a Thousand Words: The Images of Copyright," *Harvard Law Review* 125, no. 3 (2011): 683–759.

43. Julian Thomas, "The Old New Television and the New: Digital Transitions at Home," *Media International Australia* 129 (2008): 93.

44. For further work on distribution see Sean Cubitt, "Distribution and Media Flows," *Cultural Politics* 1, no. 2 (2005): 193–214, or Ramon Lobato, *Shadow Economies of Cinema: Mapping Informal Film Distribution* (London: Palgrave Macmillan, 2012).

45. Jesse Walker, *Rebels on the Air: An Alternative History of Radio in America* (New York: New York University Press, 2001), 16.

46. Burton Paulu, *Television and Radio in the United Kingdom* (Minneapolis: University of Minnesota Press, 1981), 6.

47. See Walker, *Rebels on the Air*, 20.

48. Ibid., 22. See also Radio Act of 1912, ch. 287, 37 Stat. 302 (repealed 1927).

49. Walker, *Rebels on the Air*, 29.

50. Ralph Engelman, *Public Radio and Television in America: A Political History* (California: SAGE Publications, 1996), 17.

51. Walker, *Rebels on the Air*, 30.

52. Ibid.

53. Radio Act of 1927, Pub. L. No. 69–632, 44 Stat. 1162 (1927).

54. Walker, *Rebels on the Air*, 35.

55. See Keith Negus, *Producing Pop* (London: Edward Arnold, 1992); and Matthew DelNero, "Long Overdue? An Exploration of the Status and Merit of a General Public Performance Right in Sound Recordings," *Journal of the Copyright Society of the U.S.A* 51, no. 3 (2004): 473–520.

56. See generally Jessica Litman, *Digital Copyright* (New York: Prometheus Books, 2001).

57. Paulu, *Television and Radio in the United Kingdom*, 101.

58. "The Licence Fee," BBC, accessed May 10, 2016, http://www.bbc.co.uk/about thebbc/insidethebbc/whoweare/licencefee.

59. Communications Act, 2003, C. 21, 4 (363).

60. See Paulu, *Television and Radio in the United Kingdom*, 63.

61. Thomas, "The Old New Television and the New," 93.

62. Ibid.

63. Eric Gelman et al., "The Video Revolution," *Newsweek*, August 6, 1984, 50.

64. Ibid., 57.

65. Ibid.

66. Ien Ang, *Living Room Wars: Rethinking Media Audiences for a Postmodern World* (London: Routledge, 1996), 9–10, 59–60.

67. Ibid. Also see Ien Ang, *Desperately Seeking the Audience* (London: Routledge, 1991).

68. Hilderbrand, *Inherent Vice*, 89. See also Sony Corp. of America v. Universal City Studios Inc. 464 U.S. 417 (1984).

69. Hilderbrand, *Inherent Vice*, 100.

70. Ibid., 89–91.

71. These points sit at the heart of Hilderbrand's argument in *Inherent Vice*.

72. See Yochai Benkler, *The Wealth of Networks: How Social Production Transforms Markets and Freedom* (New Haven, CT: Yale University Press, 2006); Axel Bruns, *Blogs, Wikipedia, Second Life, and Beyond: From Production to Produsage* (New York: Peter Lang, 2008); Henry Jenkins, *Convergence Culture: Where Old and New Media Collide* (New York: New York University Press, 2006); Lawrence Lessig, *Remix: Making Art and Commerce Thrive in the Hybrid Economy* (New York: Penguin, 2008); and Clay Shirky, *Here Comes Everybody: The Power of Organizing without Organizations* (New York: Penguin, 2008).

73. Benkler, *The Wealth of Networks*, 213.

74. Shirky, *Here Comes Everybody*.

75. Henry Jenkins, *Fans, Bloggers, and Gamers: Exploring Participatory Culture* (New York: New York University Press, 2006).

76. Jenkins, *Convergence Culture*, 3. This work was published over a decade ago and Jenkins has since developed more nuanced claims in response to a special *Cultural Studies* issue on this work: Henry Jenkins, "Rethinking 'Rethinking Convergence/Culture,'" *Cultural Studies* 28, no. 2 (2014): 267–297.

77. Lev Grossman, "You—Yes—You Are *TIME*'s Person of the Year," *Time*, December 25, 2006, http://content.time.com/time/magazine/article/0,9171,1570810,00.html.

78. Ibid.

79. Jean Burgess, "The iPhone Moment, the Apple Brand and the Creative Consumer: From 'Hackability and Usability' to Cultural Generativity," in *Studying Mobile*

Media: Cultural Technologies, Mobile Communication, and the iPhone, ed. Larissa Hjorth, Ingrid Richardson, and Jean Burgess (New York: Routledge, 2012), 28–42.

80. Ramon Lobato, Julian Thomas, and Dan Hunter, "Histories of User-Generated Content: Between Formal and Informal Media Economies," *International Journal of Communication* 5 (2011): 900.

81. Jason Potts et al., "Consumer Co-Creation and Situated Creativity," *Industry and Innovation* 15, no. 5 (2008): 459–474.

82. Bruns, *Blogs, Wikipedia, Second Life, and Beyond.*

83. Aufderheide and Jaszi, *Reclaiming Fair Use*, 84.

84. Warner Bros. Entertainment, Inc. and J. K. Rowling v. RDR Books, 575 F.Supp.2d 513 (SDNY 2008).

85. Shira Siskind, "Crossing the Fair Use Line: The Demise and Revival of the Harry Potter Lexicon and Its Implications for the Fair Use Doctrine in the Real World and on the Internet." *Cardozo Arts & Entertainment Law Journal* 27, no. 1 (2009): 292.

86. Warner Bros. Entertainment, Inc. and J. K. Rowling v. RDR Books, 575 F.Supp.2d 513 (SDNY 2008). Opinion and Order.

87. Aufderheide and Jaszi, *Reclaiming Fair Use*, 90.

88. Copyright Act 1968 (Cth), s41A.

89. Copyright Act, RSC 1985, c C-42, s29.21.

Chapter 3

1. Rauly Ramirez, "Robin Thicke's 'Blurred Lines' Breaks Record atop Hop R&B/Hip-Hop Songs," Billboard.com, September 25, 2013, http://www.billboard.com/articles /columns/the-juice/5733206/robin-thickes-blurred-lines-breaks-record-atop-hot -rbhip-hop.

2. A direct comparison of the two songs can be found on YouTube. See Josh Chester-field, "Robin Thicke—Blurred Lines VS Marvin Gaye—Got to Give it Up," YouTube, November 1, 2013, https://www.youtube.com/watch?v=ziz9HW2ZmmY.

3. Evan Minsker, "Marvin Gaye's Children Release Statement on 'Blurred Lines' Law-suit," Pitchfork.com, March 18, 2015, http://pitchfork.com/news/58889-marvin-gayes -children-release-statement-on-blurred-lines-lawsuit/.

4. Amy Phillips, "Jury Rules Pharrell and Robin Thicke's 'Blurred Lines' Copied Mar-vin Gaye," Pitchfork.com, March 10, 2015, http://pitchfork.com/news/58743-jury -rules-pharrell-and-robin-thickes-blurred-lines-copied-marvin-gaye/.

5. Eriq Gardner, "'Blurred Lines' Appeal Gets Support from More Than 200 Musicians," The Hollywood Reporter, August 30, 2016, http://www.hollywoodreporter .com/thr-esq/blurred-lines-appeal-gets-support-924213.

6. Eriq Gardner, "Robin Thicke Admits Drug Abuse, Lying to Media in Wild 'Blurred Lines' Deposition (Exclusive)," The Hollywood Reporter, September 15, 2014, http:// www.hollywoodreporter.com/thr-esq/robin-thicke-admits-drug-abuse-732783.

7. See Toni Lester, "Blurred Lines—Where Copyright Ends and Cultural Appropriation Begins—The Case of Robin Thicke versus Bridgeport Music and the Estate of Marvin Gaye," *Hastings Communications and Entertainment Law Journal* 36, no. 2 (2013): 217–242.

8. See Carys Craig, *Copyright, Communication and Culture: Towards a Relational Theory of Copyright Law* (Cheltenham, UK: Edward Elgar Publishing, 2011), in particular chapter 3.

9. See William Klenz, "Brahms Op. 38; Piracy, Pillage, Plagiarism or Parody?" *Music Review* 34, no. 1 (1973): 39–50; and Jeremy Yudkin, "Beethoven's 'Mozart' Quartet." *Journal of the American Musicological Society* 45, no. 1 (1992): 30–74.

10. Kembrew McLeod and Peter DiCola, *Creative License: The Law and Culture of Digital Sampling* (Durham, NC: Duke University Press, 2011), 47.

11. See "Jazz Has Got Copyright Law and That Ain't Good," *Harvard Law Review* 118, no. 6 (2005): 1940–1961.

12. See Stephen Wilson, "Rewarding Creativity: Transformative Use in the Jazz Idiom," *Pittsburgh Journal of Technology Law and Policy* 4 (2003–2004): 1–32; and Olufunmilayo Arewa, "From JC Bach to Hip Hop: Musical Borrowing, Copyright and Cultural Context," *North Carolina Law Review* 84, no. 2 (2006): 547–645.

13. McLeod and DiCola recount this history in detail in their work *Creative License*. Johnson Okpaluba also offers a history of sampling in the following study: "Digital Sampling and Music Industry Practices, Re-Spun," in *Law and Creativity in the Age of the Entertainment Franchise*, ed. Kathy Bowrey and Michael Handler (Cambridge, UK: Cambridge University Press, 2014), 75–102.

14. Kembrew McLeod, "How Copyright Law Changed Hip Hop," Alternet, May 31, 2004, http://www.alternet.org/story/18830/how_copyright_law_changed_hip_hop.

15. Dorian Lynskey, "My Favourite Album: It Takes a Nation of Millions to Hold Us Back by Public Enemy." *The Guardian*, August 11, 2011, http://www.theguardian.com /music/musicblog/2011/aug/11/public-enemy-nation-millions-hold-us-back.

16. McLeod and DiCola, *Creative License*, 28.

17. Ibid., 26.

18. Ibid., 27.

19. Grand Upright Music Ltd. v. Warner Bros. Records, Inc. 780 F. Supp. 182 (S.D.N.Y. 1991).

20. John Schietinger, "Bridgeport Music, Inc. v. Dimension Films: How the Sixth Circuit Missed a Beat on Digital Music Sampling," *DePaul Law Review* 55, no. 1 (2005): 222.

21. Campbell v. Acuff-Rose Music, 510 U.S. 569 (1994).

22. See McLeod and DiCola, *Creative License*.

23. Okpaluba, "Digital Sampling and Music Industry Practices, Re-Spun," 83–84.

24. Ibid., 97, 92.

25. For a detailed description of these processes, see McLeod and DiCola, *Creative License*, 154.

26. Ibid., 158.

27. See Bridgeport Music, Inc. v. Dimension Films, 230 F. Supp. 2d 830 (M.D. Tenn. 2002); Bridgeport Music, Inc. v. Dimension Films 410 F.3d 792 (6th Cir. 2005).

28. Schietinger, "Bridgeport Music, Inc. v. Dimension Films," 224.

29. See Bridgeport Music, Inc. v. Dimension Films, 230 F. Supp. 2d 830 (M.D. Tenn. 2002).

30. Ibid., at 839.

31. Bridgeport Music, Inc. v. Dimension Films, 383 F.3d 390, 398 (6th Cir. 2004).

32. Ibid., at 87.

33. Andrew Currah discusses how Hollywood attempts to control innovation in order to produce an organized and predictable marketplace in "Hollywood, the Internet and the World: A Geography of Disruptive Innovation," *Industry and Innovation* 14, no. 4 (2007): 359–384.

34. Kembrew McLeod, *Freedom of Expression: Overzealous Copyright Bozos and Other Enemies of Creativity* (New York: Doubleday, 2005); and McLeod and DiCola, *Creative License*.

35. Craig, *Copyright, Communication and Culture*, 221.

36. Arewa, "From JC Bach to Hip Hop," 556–557.

37. Larrikin Music Publishing Pty Ltd. v. EMI Songs Australia Pty Ltd. [2010] FCA 29.

38. Hi-5, "Hi-5 Charli on Spicks 'n Specks," YouTube, filmed November 2007, posted November 20, 2007, https://www.youtube.com/watch?v=IT8SHafGIpU,quote at minute 1:50.

39. Phoebe Vertigan, "EMI Songs Australia Pty Ltd v Larrikin Music Publishing Pty Ltd (2011) 191 FCR 444," *University of Tasmania Law Review* 30, no. 2 (2011): 155.

40. Larrikin Music Publishing Pty Ltd. v. EMI Songs Australia Pty Ltd. [2010] FCA 29.

41. EMI Songs Australia Pty Ltd. v. Larrikin Music Publishing Pty Ltd. [2011] FCAFC 47. There were a number of cases in this dispute that I have not discussed in the text that involved adjudications around who owned the copyrighted work and the issue of damages. For a detailed background to this longer case history, see Matthew Rimmer, "An Elegy for Greg Ham: Copyright Law, the Kookaburra Case, and Remix Culture." *Deakin Law Review* 17, no. 2 (2013): 387.

42. EMI Songs Australia Pty Ltd. v. Larrikin Music Publishing Pty Ltd. [2011] FCAFC 47 at 198–199.

43. Sarah Dingle, "Kookaburra Rip-Off a 'Musical Accident,'" ABC News, last updated February 4, 2010, http://www.abc.net.au/news/2010-02-04/kookaburra-rip-off-a -musical-accident/322496.

44. See Rebecca Tushnet, "Worth a Thousand Words: The Images of Copyright," *Harvard Law Review* 125, no. 3 (2011):716–717.

45. Ibid., 740.

46. Ibid., 737.

47. Ibid., 711.

48. Ibid., 739.

49. "Police Investigate Death of Men at Work Flautist," ABC News, last updated April 19, 2012, http://www.abc.net.au/news/2012-04-19/men-at-works-greg-ham-found -dead/3960780.

50. Ibid. Sadly, Greg Ham, the flautist for Men at Work who wrote the infamous solo, died on April 19, 2012, soon after the case ended. See also Rimmer, "An Elegy for Greg Ham."

51. Stuart Jeffries, "The Chapman Brothers on Life as Artists' Assistants: 'We Did Our Daily Penance,'" *The Guardian*, March 23, 2013, http://www.theguardian.com /artanddesign/2013/mar/23/artists-assistants-chapman-brothers.

52. Anita Singh, "Damien Hirst: Assistants Make My Spot Paintings but My Heart Is in Them All," *Telegraph*, January 12, 2012, http://www.telegraph.co.uk/culture/art /art-news/9010657/Damien-Hirst-assistants-make-my-spot-paintings-but-my-heart -is-in-them-all.html.

53. Ibid.

54. On occasion, architects have been granted copyright in works that assistants have produced. See Lionel Bently and Laura Biron, "Discontinuities between Legal

Conceptions of Authorship and Social Practices," in *The Work of Authorship*, ed. Mireille van Eechoud (Amsterdam: University of Amsterdam Press, 2014), 247.

55. See 17 U.S.C. § 201(b) and generally Catherine Fisk, "Authors at Work: The Origins of the Work-for-Hire Doctrine," *Yale Journal of Law & the Humanities* 15, no. 1 (2013): 1–70.

56. See Copyright Act 1968 (Cth) s35(6); Copyright Act, RSC 1985, c C-42 13(3); and Copyright, Designs and Patents Act 1988, c. 48 (UK), s11(2).

57. Copyright Act, RSC 1985, c C-42 17.1(2); Copyright, Designs and Patents Act 1988, c. 48 (UK), s87; and 17 U.S.C. § 106A(e).

58. Copyright Act 1968 (Cth) s195AW, s195AWA.

59. Xi Yin Tang, "The Artist as Brand: Toward a Trademark Conception of Moral Rights," *Yale Law Journal* 122, no. 1 (2012): 234.

60. See Bently and Biron, "Discontinuities between Legal Conceptions of Authorship and Social Practices."

61. Amy Adler, "Against Moral Rights," *California Law Review* 91, no. 1 (2009): 263–300.

62. Henry Hansmann and Marina Santilli, "Authors' and Artists' Moral Rights: A Comparative Legal and Economic Analysis," *Journal of Legal Studies* 26, no. 1 (1997): 95–96.

63. Adler, "Against Moral Rights," 296.

64. Tang, "The artist as Brand," 234.

65. Ibid., 239.

66. Ibid., 248.

67. Ibid.

68. Jeffries, "The Chapman Brothers."

69. Daniel Grant, "Low Pay, Monotonous Work: Are Artist Assistant Positions Worth the Trouble?," Observer, July 22 2014, http://observer.com/2014/07/low-pay-monotonous-work-are-artist-assistant-positions-worth-the-trouble/. See also Brian Sherwin, "Artists Debate Over the Use of Artist Assistants–Where Do You Stand?," *Faso* (blog), January 8, 2012, http://faso.com/fineartviews/38751/artists-debate-over-the-use-of-artist-assistants-where-do-you-stand.

70. See Kathy Bowrey and Michael Handler, "Instituting Copyright: Reconciling Copyright Law and Industry Practice in the Australian Film and Television Sector," in *Law and Creativity in the Age of the Entertainment Franchise*, ed. Kathy Bowrey and Michael Handler (Cambridge, UK: Cambridge University Press, 2014), 140–169; and

Catherine Fisk, "Will Work for Screen Credit: Labour and the Law in Hollywood," in *Hollywood and the Law*, ed. Paul McDonald et al. (London: Palgrave, 2015), 235.

71. Bowrey and Handler, "Instituting Copyright."

72. Tang, "The Artist as Brand."

73. Laura Heymann, "Everything Is Transformative: Fair Use and Reader Response," *Columbia Journal of Law and the Arts* 31, no. 4 (2008): 460. See also Rogers v. Koons, 960 F.2d 301, 304 (2d Cir. 1992).

74. United Features Syndicate, Inc. v. Koons, 817 F. Supp. 370 (S.D.N.Y. 1993).

75. Blanch v. Koons, 467 F.3d 244 (2d Cir. 2006) at 16.

76. Ibid., at 13.

77. Ibid., at 21.

78. Ibid., at 17.

79. Blanch v. Koons, 396 F.Supp.2d 476 (S.D.N.Y. 2005).

80. Blanch v. Koons, 467 F.3d 244 (2d Cir. 2006).

81. Ibid., at 66, 54.

82. See McLeod, *Freedom of Expression*; and McLeod and DiCola, *Creative License*. Lawrence Lessig also discusses the broader issues that come with establishing a culture of permission around copyrighted works in *Free Culture: How Big Media Uses Technology and the Law to Lock Down Culture and Control Creativity* (New York: Penguin Press, 2004).

83. See Kathy Bowrey, "Copyright, the Paternity of Artistic Works, and the Challenge Posed by Postmodern Artists," *Intellectual Property Journal* 8, no. 3 (1994): 285–317.

84. See Campbell v. Acuff-Rose Music, Inc., 510 U.S. 577 (1994).

85. Olav Velthuis, "Symbolic Meanings of Prices: Constructing the Value of Contemporary Art in Amsterdam and New York Galleries," *Theory and Society* 32, no. 2 (2003): 193.

86. See Robert Hughes, *The Shock of the New* (New York: Knopf, 1981).

87. Amy Adler, "Fair Use and the Future of Art," *New York University Law Review* 91, no. 3 (2016): 559–626.

88. Ibid.

89. Ibid., 582.

90. Ibid., 583.

91. Ibid., 592, 593.

92. A similar argument is advanced by Bowrey and Handler in "Instituting Copyright."

Chapter 4

1. For critical approaches to the broad conceptualization of the user subject in law and media studies, see: Julie Cohen, "The Place of the User in Copyright Law," *Fordham Law Review* 74 (2005): 347–374; Johanna Gibson, *Creating Selves: Intellectual Property and the Narration of Culture* (Aldershot, UK: Ashgate Publishing, 2006); Bridget Griffen-Foley, "From Tit-Bits to Big Brother: A Century of Audience Participation in the Media," *Media, Culture & Society* 26, no. 4 (2004): 533–548; James Hamilton, "Historical Forms of User Production," *Media, Culture & Society* 36, no. 4 (2014): 491–507; Joseph Liu, "Copyright Law's Theory of the Consumer," *Boston College Law Review* 44, no. 2 (2003): 397–431; James Meese, "User Production and Law Reform: A Socio-Legal Critique of User Creativity," *Media, Culture & Society* 37, no. 5 (2015): 753–767; and Kylie Pappalardo, "A Tort Law Framework for Copyright Authorisation" (PhD thesis, Australian Catholic University, 2015).

2. See Yochai Benkler, *The Wealth of Networks: How Social Production Transforms Markets and Freedom* (New Haven, CT: Yale University Press, 2006); Dan Hunter et al., eds., *Amateur Media: Social, Cultural and Legal Perspectives* (Oxford, UK: Routledge, 2013); and Clay Shirky, *Here Comes Everybody: The Power of Organizing without Organizations* (New York: Penguin, 2008).

3. Cohen, "The Place of the User," 374.

4. Ibid.

5. Pappalardo, "A Tort Law Framework for Copyright Authorisation."

6. For recent studies on how copyright intersects with the galleries, libraries, archives, and museums (GLAM) sector, see Emily Hudson and Andrew Kenyon, "Digital Access: The Impact of Copyright on Digitisation Practices in Australian Museums, Galleries, Libraries, and Archives," *University of New South Wales Law Journal* 30, no. 1 (2007): 12–52; and Peter Hirtle, Emily Hudson, and Andrew Kenyon. *Copyright and Cultural Institutions: Guidelines for Digitizations for U.S. Libraries, Archives and Museums* (Ithaca: Cornell University Library Press, 2009).

7. Meese, "User Production and Law Reform," 754. See also Hamilton, "Historical Forms of User Production."

8. Copyright Amendment Act 2006 (Cth).

9. The Senate of Australia, *Select Committee on the Free Trade Agreement between Australia and the United States of America: Final Report* (Canberra, Parliament House, 2004), 77.

10. Copyright Act 1968 (Cth), s40–44.

11. See Lawrence Lessig, *Remix: Making Art and Commerce Thrive in the Hybrid Economy* (New York: Penguin Press, 2008); Damien O'Brien and Brian Fitzgerald, "Mashups, Remixes and Copyright Law," *Internet Law Bulletin* 9, no. 2 (2006): 17–19; Damien O'Brien and Brian Fitzgerald, "Digital Copyright Law in a YouTube World," *Internet Law Bulletin* 9, no. 6 & 7 (2006): 71–74.

12. Commonwealth Attorney General and Phillip Ruddock, *Major Copyright Reforms Strike Balance*, press release, 2006.

13. Copyright Act 1968 (Cth), s111.

14. Ibid., s41A.

15. The legal concept of strict liability means that individuals can be made responsible for their actions regardless of their intent. In this case, it would mean that individuals would be liable for acts that infringed copyright even if they were unaware they were doing so and had no intent to infringe.These "strict liability" offenses were introduced into a number of sections in the act. For relevant discussions of their introduction, see: Catherine Bond, Abi Paramaguru, and Graham Greenleaf, "Advance Australia Fair? The Copyright Reform Process." *Journal of World Intellectual Property* 10, no. 3–4 (2007): 284–313; or Kimberlee Weatherall, "Of Copyright Bureaucracies and Incoherence: Stepping Back from Australia's Recent Copyright Reforms," *Melbourne University Law Review* 31, no. 3 (2007): 967–1174.

16. Ian Hargreaves, *Digital Opportunity: A Review of Intellectual Property and Growth* (London: Department for Business, Innovation and Skills, 2011), 41.

17. Ibid., 49.

18. Ibid.

19. Ibid., 50.

20. See *Consultation on Copyright: Summary of Responses June 2012*, (London: Intellectual Property Office, 2012), accessed May 21, 2016, https://www.gov.uk/govern ment/uploads/system/uploads/attachment_data/file/320223/copyright-consultation -summary-of-responses.pdf.

21. Ibid., 19.

22. Intellectual Property Office, "Changes to Copyright Exceptions," Gov.uk, July 30, 2014, https://www.gov.uk/government/news/changes-to-copyright-exceptions.

23. HL Deb 19 July 2014, vol. 755, col. 1564. See http://www.publications.parliament .uk/pa/ld201415/ldhansrd/text/140729-0002.htm.

24. UK Music, *Press Notice*, press release, accessed May 21, 2016, http://www.ukmusic .org/news/press-notice.

25. Intellectual Property Office, "Quashing of Private Copying Exception," Gov.uk, July 20 2015, https://www.gov.uk/government/news/quashing-of-private-copying -exception.

26. Copyright Act, RSC 1985, c C-42 amended by the Copyright Modernization Act 2012.

27. Meese, "User Production and Law Reform," 761. See generally Michael Geist, ed., *The Copyright Pentalogy* (Ottawa: University of Ottawa Press, 2013).

28. Copyright Act, RSC 1985, c C-42, s29.21.

29. Ibid.

30. Michael Geist, "What the New Copyright Law Means for You," MichaelGeist.ca, November 13, 2012, http://www.michaelgeist.ca/2012/11/c-11-impact/.

31. Berne Convention for the Protection of Literary and Artistic Works, Art. 9(2). The test is contested and has been examined in detail with scholars debating the best way to interpret the "open-ended" wording of the condition for particular national jurisdictions. See Christophe Geiger, Daniel Gervais, and Martin Senftleben, "The Three-Step Test Revisited: How to Use the Test's Flexibility in National Copyright Law," *American University International Law Review* 29, no. 3 (2014): 581–626; or Christophe Geiger, "From Berne to National Law, via the Copyright Directive: The Dangerous Mutations of the Three-Step Test," *European Intellectual Property Review* 29, no. 12 (2007): 486–491.

32. Kamiel Koelman, "Fixing the Three-Step Test," *European Intellectual Property Review* 28, no. 8 (2006): 408.

33. Peter Yu, "Can the Canadian UGC Exception Be Transplanted Abroad?," *Intellectual Property Journal* 26, no. 2 (2014): 175–203.

34. Teresa Scassa, "Acknowledging Copyright's Illegitimate Offspring: User-Generated Content and Canadian Copyright Law," in *The Copyright Pentalogy: How the Supreme Court of Canada Shook the Foundations of Canadian Copyright Law*, ed. Michael Geist (Ottawa, CA: University of Ottawa Press, 2013), 437.

35. Meese, "User Production and Law Reform," 762.

36. Copyright Act, RSC 1985, c C-42, s29.21(1)(d).

37. Meese, "User Production and Law Reform," 763.

38. Scassa, "Acknowledging Copyright's Illegitimate Offspring," 443.

39. See Meese, "User Production and Law Reform," or Kristofer Erickson, "User Illusion: Ideological Construction of 'User-Generated Content' in the EC Consultation on Copyright," *Internet Policy Review* 3, no. 4 (2014): 1–19.

40. See Erickson, "User Illusion."

41. Ibid., 10.

42. Lionel Maurel and Philippe Aigrain, *La Quadrature du Net's Response to the European Commission's Consultation on Copyright Reform*, February 2014, 20, available at https://www.laquadrature.net/files/La_Quadrature_du_Net_s_response_to_the_European_Commission_s_consultation_on_copyright_reform.pdf.

43. Ibid.

44. Erickson, "User Illusion," 11.

45. Ibid., 12.

46. For a detailed discussion of how the consumer was framed during the UK reform process, see Lee Edwards et al., "Framing the Consumer: Copyright Regulation and the Public," *Convergence: The International Journal of Research into New Media Technologies* 19, no. 1 (2013): 9–24. An analysis of how the user was situated throughout the Australian reform process can be found in James Meese, "Networked Subjects: Exploring the Author, User and Pirate through a Relationalist Lens" (PhD thesis, Swinburne University of Technology, 2014).

47. Erickson, "User Illusion," 11.

48. Ibid.

49. See generally Jean Burgess and Joshua Green, *YouTube: Online Video and Participatory Culture* (Cambridge, UK: Polity Press, 2012); Pelle Snickars and Patrick Vonderau eds., *The YouTube Reader* (Stockholm: National Library of Sweden, 2009); Joanne Morreale, "From Homemade to Store Bought: Annoying Orange and the Professionalization of YouTube," *Journal of Consumer Culture* 14, no. 1 (2014): 113–128; Jin Kim, "The Institutionalization of YouTube: From User-Generated Content to Professionally Generated Content," *Media, Culture & Society* 34, no. 1 (2012): 53–67; and Maura Edmond, "Here We Go Again: Music Videos after YouTube," *Television & New Media* 15, no. 4 (2014): 305–320.

50. Ramon Lobato, "The Cultural Logic of Digital Intermediaries: YouTube Multichannel Networks," *Convergence: The International Journal of Research into New Media Technologies* 22, no. 4 (2016): 348–360. See also Patrick Vonderau, "The Video Bubble: Multichannel Networks and the Transformation of YouTube," *Convergence: The International Journal of Research into New Media Technologies* 22, no. 4 (2016): 361–375.

51. Lobato, "The Cultural Logic of Digital Intermediaries," 352.

52. Ibid.

53. It was recently revealed that PewDiePie featured anti-Semitic content in some of his videos. He was dropped from the Disney-owned MCN Maker Studios and YouTube canceled his YouTube Red series. He is still able to make money through standard monetization of his YouTube content.

54. Stephanie Mlot, "Nintendo Placing Ads on Users' Youtube Gameplay Videos," PCMag, May 17, 2013, http://www.pcmag.com/article2/0,2817,2419134,00.asp.

55. This program is found at https://r.ncp.nintendo.net/.

56. Patricia Hernandez, "Nintendo's YouTube Plan Is Already Being Panned by You-Tubers," Kotaku, January 30, 2015, http://www.kotaku.com.au/2015/01/nintendos-youtube-plan-is-already-being-panned-by-youtubers/.

57. Chris Kohler, "Why Does Nintendo Want This Superfan's YouTube Money?" Wired, March 27, 2015, http://www.wired.com/2015/03/nintendo-youtube-creators/.

58. "How Content ID Works," YouTube Help, accessed May 21, 2016, https://support.google.com/youtube/answer/2797370?hl=en.

59. Kohler, "Why Does Nintendo Want This Superfan's YouTube Money?"

60. Tarleton Gillespie, *Wired Shut: Copyright and the Shape of Digital Culture* (Cambridge, MA: MIT Press, 2007), 8.

61. David King, "Latest Content ID Tool for YouTube," Google's Official Blog, October 15, 2007, https://googleblog.blogspot.com.au/2007/10/latest-content-id-tool-for-youtube.html.

62. See Fred Von Lohmann, "YouTube's January Fair Use Massacre," Electronic Frontier Foundation, February 3, 2009, https://www.eff.org/deeplinks/2009/01/youtubes-january-fair-use-massacre.

63. Saba Hamedy, "YouTube Takes Step Toward Protecting Creators from Copyright Claims," Mashable, November 19, 2015, http://mashable.com/2015/11/19/youtube-takes-step-toward-protecting-fair-use/#RFQqJwi3n5q2.

64. This argument was advanced in industry blog *Tech Crunch*: David Felicissimo, "Google's Move Toward Fair Use Comes in Anticipation of YouTube Red," *TechCrunch* (blog), December 17, 2015, https://techcrunch.com/2015/12/17/googles-move-toward-fair-use-in-anticipation-of-youtube-red/.

65. Ibid.

66. 2 17 U.S.C. § 101 (2012).

67. Rebecca Giblin and Jane Ginsburg, "We (Still) Need to Talk about Aereo: New Controversies and Unresolved Questions after the Supreme Court's Decision," *Columbia Journal of Law & the Arts* 38, no. 2 (2015): 117–118.

68. Ibid., 109.

69. Optus v. NRL [2012] FCAFC 59 at 51, 34.

70. Rebecca Giblin, "Optus v NRL: A Seismic Shift for Time Shifting in Australia," *European Intellectual Property Review* 34, no. 5 (2012): 358.

71. This case culminated at the Supreme Court in 2014 (Am. Broad. Cos., Inc. v. Aereo, Inc., 134 S. Ct. 2498).

72. Copyright Act 1968 (Cth) s202.

73. Giblin, "Optus v NRL," 358.

74. Optus v NRL [2012] FCA 34 at 63.

75. See Giblin and Ginsburg, "We (Still) Need to Talk about Aereo," 110.

76. Optus v NRL [2012] FCA 34 at 65.

77. Ibid., at 112, 68.

78. Cartoon Network LP v. CSC Holdings, Inc., 536 F.3d 121, (2d Cir. 2008).

79. Am. Broad. Cos. v. Aereo, Inc. 874 F.Supp. 2d 373, 385 (S.D.N.Y. 2012).

80. Ibid., at 396.

81. Optus v NRL [2012] FCAFC 59 at 67.

82. Ibid., at 89.

83. See WNET, Thirteen v. Aereo, Inc., 712 F.3d 676, (2d Cir. 2013).

84. Am. Broad. Cos., Inc. v. Aereo, Inc., 134 S. Ct. 2498 (2014). See also John Eggerton, "Supreme Court to Hear Aereo Appeal," Broadcasting & Cable, January 10, 2014, http://www.broadcastingcable.com/news/washington/supreme-court-hear-aereo -appeal/128413.

85. Am. Broad. Cos., Inc. v. Aereo, Inc., 134 S. Ct. 2498 (2014) at 54.

86. Ibid., at 60.

87. Ibid., at 68.

88. Giblin, "Optus v NRL."

89. See Brief of 36 Intellectual Property and Copyright Law Professors as *Amici Curae* in Support of Respondent, Am. Broad. Cos., Inc. v. Aereo, Inc., S. Ct.

90. CCH Canadian Ltd. v. Law Society of Upper Canada, [2004] 1 S.C.R. 339, 2004 SCC 13.

91. Ibid., at 5.

92. University of NSW v Moorhouse (1975) HCA 26.

93. CCH Canadian Ltd. v. Law Society of Upper Canada, at 45.

94. Copyright Act, RSC 1985, c C-42, s29.

95. CCH Canadian Ltd. v. Law Society of Upper Canada, at 64.

96. Society of Composers, Authors and Music Publishers of Canada v. Bell Canada, 2012 SCC 36, [2012] 2 S.C.R. 326.

97. Ibid., at 19.

98. Ibid., at 29.

99. Ibid., at 34.

100. Alberta (Education) v. Canadian Copyright Licensing Agency (Access Copyright), 2012 SCC 37, [2012] 2 S.C.R 345.

101. Ibid., at 7.

102. Ibid., at 23.

103. Cohen, "The Place of the User," 348.

104. Ibid., 348–349.

105. Ibid., 370.

106. Julie Cohen, "Creativity and Culture in Copyright Theory," *U.C. Davis Law Review* 40, no. 3 (2007): 1179.

107. See generally, Pappalardo, "A Tort Law Framework for Copyright Authorisation."

108. Ibid., 117–129.

109. Ibid., 129–132.

110. Cohen, "The Place of the User," 374.

111. See Meese, "User Production and Law Reform"; and Hamilton, "Historical Forms of User Production."

112. Maria Bakardjieva, *Internet Society: The Internet in Everyday Life* (London: Sage Publications, 2005).

113. Steve Woolgar, "Configuring the User: the Case of Usability Trials," *Sociological Review* 38, no. S1 (1990): 69.

114. Bakardjieva, *Internet Society*, 13.

115. In addition to the literature discussed previously, see also Madeline Akrich, "The De-Scription of Technical Objects" in *Shaping Technology, Building Society: Studies in Sociotechnical Change*, ed. Wiebe Bijker and John Law (Cambridge, MA: MIT Press, 1992): 205–224; Graeme Kirkpatrick, *Critical Technology: A Social Theory of Personal Computing* (Aldershot, UK: Ashgate, 2004); and Bruno Latour, "Where Are the Missing Masses? Sociology of a Few Mundane Artifacts," in *Shaping Technology, Building Society: Studies in Sociotechnical Change*, ed. Wiebe Bijker and John Law (Cambridge, MA: MIT Press, 1992): 225–258.

116. Woolgar, "Configuring the User."

117. Bakardjieva, *Internet Society*, 21.

Chapter 5

1. John Logie, *Peers, Pirates, and Persuasion: Rhetoric in the Peer-to-Peer Debates* (West Lafayette, IN: Parlor Press, 2006), 53. This is partly because copyright law only legally recognizes (and enforces) the infringement of copyright rather than the highly emotive concept of "piracy." See also Jessica Reyman, *The Rhetoric of Intellectual Property* (New York: Routledge, 2010); Patricia Loughlan, "Pirates, Parasites, Reapers, Sowers, Fruits, Foxes … The Metaphors of Intellectual Property," *Sydney Law Review* 28, no. 2 (2006): 211–226; and Patricia Loughlan, "'You Wouldn't Steal a Car': Intellectual Property and the Language of Theft," *European Intellectual Property Review* 29, no. 10 (2007): 401–405.

2. Stop Online Piracy Act, H.R. 3261, 112th. Cong. (2011) (hereafter SOPA).

3. PROTECT IP Act, S. 968, 112th Cong. (2011) (hereafter PIPA).

4. Rebecca Tushnet, "Copy This Essay: How Fair Use Doctrine Harms Free Speech and How Copying Serves It," *Yale Law Journal* 114 (2004): 535–590; and Kal Raustiala and Christopher Sprigman, *The Knockoff Economy: How Imitation Sparks Innovation* (Oxford, UK: Oxford University Press, 2012).

5. Lawrence Lessig, *Remix: Making Art and Commerce Thrive in the Hybrid Economy* (New York: Penguin, 2008), 27.

6. Ibid., 38.

7. Anupam Chander and Madhavi Sunder, "Everyone's a Superhero: A Cultural Theory of 'Mary Sue' Fan Fiction as Fair Use," *California Law Review* 95, no. 2 (2007): 600.

8. Lawrence Lessig, *Free Culture: How Big Media Uses Technology and the Law to Lock Down Culture and Control Creativity* (New York: Penguin Press, 2004), 63.

9. Lucas Hilderbrand, *Inherent Vice: Bootleg Histories of Videotape and Copyright* (Durham, NC: Duke University Press, 2009), 23.

10. Ibid.

11. Adrian Johns, *Piracy: The Intellectual Property Wars from Gutenberg to Gates* (Chicago: University of Chicago Press, 2009), 6.

12. Tushnet, "Copy This Essay."

13. Ibid., 537.

14. Ibid., 567.

15. Kavita Philip, "What Is a Technological Author? The Pirate Function and Intellectual Property," *Postcolonial Studies* 8, no. 2 (2005), 199, 212.

16. Ibid., 207.

17. Lawrence Liang, "Beyond Representation: The Figure of the Pirate," in *Access to Knowledge in the Age of Intellectual Property*, ed. Gaëlle Krikorian and Amy Kapczynski (Cambridge, MA: MIT Press, 2010), 357.

18. Ibid.

19. Martin Fredriksson, "The Pirate Party and the Politics of Communication," *International Journal of Communication* 9 (2015): 909–924. See also Martin Fredriksson, "Copyright Culture and Pirate Politics," *Cultural Studies* 28, no. 5–6 (2014): 1022–1047; and Martin Fredriksson,"Piracy, Globalisation and the Colonisation of the Commons," *Global Media Journal: Australian Edition* 6, no. 1 (2012), available at http://www.hca.westernsydney.edu.au/gmjau/archive/v6_2012_1/pdf/martin_fredriks son_RA_6_1_2012.pdf.

20. Fredriksson, "The Pirate Party and the Politics of Communication."

21. See Lawrence Lessig, *Republic, Lost: How Money Corrupts Congress—and a Plan to Stop It* (New York: Twelve, 2011).

22. Patrick Burkart, *Pirate Politics* (Cambridge, MA: The MIT Press, 2014), 31.

23. Philip, "What Is a Technological Author?"

24. James Meese, "The Pirate Imaginary and the Potential of the Authorial Pirate," in *Piracy: Leakages from Modernity*, ed. Martin Fredriksson and James Avanitakis (Sacramento, CA: Litwin Books, 2014), 23. See also Bill Herman, *The Fight over Digital Rights: The Politics of Copyright and Technology* (Cambridge, UK: Cambridge University Press, 2013).

25. SOPA §103(a)(1)(B).

26. Ibid., §102(c)(2)(A)(i).

27. Declan McCullagh, "SOPA's Latest Threat: IP Blocking, Privacy-Busting Packet Inspection," CNET, November 19, 2011, http://www.cnet.com/au/news/sopas-latest -threat-ip-blocking-privacy-busting-packet-inspection/.

28. Michael Carrier, "SOPA, PIPA, ACTA, TPP: An Alphabet Soup of Innovation-Stifling Copyright Legislation and Agreements," *Northwestern Journal of Technology and Intellectual Property* 11, no. 2 (2013): 22.

29. Edward Lee, *The Fight for the Future: How People Defeated Hollywood and Saved the Internet—For Now* (Self-Published, 2013). Available at http://thefightforthefuture.com /wp-content/uploads/2013/07/The-Fight-for-the-Future_-How-P-Lee-Edward.pdf.

30. Carrier, "SOPA, PIPA, ACTA, TPP," 22.

31. Hayley Tsukayama, Beth Marlowe, and Dominic Basulto, "SOPA (Stop Online Piracy Act) Lawmaker Opposition Grows as Debate Heats Up," *Washington Post*, November 19, 2011, https://www.washingtonpost.com/business/sopa-stop-online -piracy-act-lawmaker-opposition-grows-as-debate-heats-up/2011/11/18/gIQADB dQZN_story.html.

32. Lee, *The Fight for the Future*.

33. Ibid.

34. Ibid.

35. Ibid.

36. Ibid.

37. See ibid., as well as Meese, "The Pirate Imaginary and the Potential of the Authorial Pirate."

38. Jenna Wortham, "With Twitter, Blackouts and Demonstrations, Web Flexes Its Muscle," *New York Times*, January 18, 2012, http://www.nytimes.com/2012/01/19 /technology/protests-of-antipiracy-bills-unite-web.html.

39. Hayley Tsukayama, Greg Sargent, and Sarah Halzack, "SOPA Bill Shelved after Global Protests from Google, Wikipedia and Others," *Washington Post*, January 20, 2012, https://www.washingtonpost.com/business/economy/sopa-bill-shelved-after -global-protests-from-google-wikipedia-and-others/2012/01/20/gIQAN5JdEQ_story .html.

40. Lamar Smith, "Setting the Record Straight on SOPA," *The Hill*, December 14, 2011, http://thehill.com/blogs/congress-blog/technology/199385-setting-the-record -straight-on-sopa.

41. PIPA S. 968, 112th Cong. (2011), Sec. 3.

42. Hayley Tsukayama, "Google to State Anti-SOPA Stance on Home Page," *Washington Post*, January 17, 2012, https://www.washingtonpost.com/pb/business/tech-nology/google-to-state-anti-sopa-stance-on-home-page/2012/01/17/gIQANeD05P _story.html.

43. Sue Gardner, "English Wikipedia Anti-SOPA Blackout," Wikimedia Foundation, January 16, 2012, https://wikimediafoundation.org/wiki/English_Wikipedia_anti-SOPA _blackout.

44. "Wikipedia:SOPA Initiative/Action," *Wikipedia*, accessed May 23, 2016, https:// en.wikipedia.org/wiki/Wikipedia:SOPA_initiative/Action.

45. Ramon Lobato, "Constructing the Pirate Audience: On Popular Copyright Critique, Free Culture and Cyber-Libertarianism," *Media International Australia* 139, no. 1 (2011): 121.

46. Kal Raustiala and Christopher Sprigman, "The Piracy Paradox: Innovation and Intellectual Property in Fashion Design," *Virginia Law Review* 92, no. 8 (2006): 1687–1777.

47. Dotan Oliar and Christopher Sprigman, "There's No Free Laugh (Anymore): The Emergence of Intellectual Property Norms and the Transformation of Stand-Up Comedy," *Virginia Law Review* 94, no. 8 (2008): 1787–1867.

48. Emmanuelle Fauchart and Eric Von Hippel, "Norms-Based Intellectual Property Systems: The Case of French Chefs," *Organization Science* 19, no. 2 (2008): 187–201.

49. Kate Darling, "IP without IP: A Study of the Online Adult Entertainment Industry," *Stanford Technology Law Review* 17, no. 3 (2013): 709–771.

50. Raustiala and Sprigman, *The Knockoff Economy*, 43.

51. Ibid., 71.

52. Ibid., 75.

53. Ibid., 90.

54. For work on the history of copying and the concepts and values that underpin this notion, see Marcus Boon, *In Praise of Copying* (Cambridge, MA: Harvard University Press, 2011); see also Margie Borschke, *This Is Not a Remix: Piracy, Authenticity and Popular Music* (London: Bloomsbury Publishing, 2017).

55. See Tim Highfield, *Social Media and Everyday Politics* (Malden, MA: Polity, 2016); and James Meese, "'It Belongs to the Internet': Animal Images, Attribution Norms and the Politics of Amateur Media Production," *M/C Journal* 17, no. 2 (2014), available at http://journal.media-culture.org.au/index.php/mcjournal/article/view/782.

56. Jenny Kennedy, "Conceptual Boundaries of Sharing," *Information, Communication & Society* 19, no. 4 (2016): 463.

57. See Nicholas John, "File Sharing and the History of Computing: or, Why File Sharing Is Called 'File Sharing,'" *Critical Studies in Media Communication* 31, no. 3 (2014): 198–211.

58. Jenny Kennedy, "Rhetorics of Sharing: Data, Imagination and Desire," in *Unlike Us Reader: Social Media Monopolies and Their Alternatives*, ed. Geert Lovink and Miriam Rasch (Amsterdam: Institute of Network Cultures, 2013), 127–136.

59. Ibid. See also Jean Burgess, "From 'Broadcast Yourself' to 'Follow Your Interests': Making Over Social Media," *International Journal of Cultural Studies* 18, no. 3 (2015): 281–285.

60. See Ramon Lobato, "The Cultural Logic of Digital Intermediaries: YouTube Multichannel Networks," *Convergence: The International Journal of Research into New Media*

Technologies 22, no. 4 (2016): 348–360; Joanne Morreale, "From Homemade to Store Bought: Annoying Orange and the Professionalization of YouTube," *Journal of Consumer Culture* 14, no. 1 (2014): 113–128; and Patrick Vonderau, "The Video Bubble: Multichannel Networks and the Transformation of YouTube," *Convergence: The International Journal of Research into New Media Technologies* 22, no. 4 (2016): 361–375.

61. Henry Jenkins, Sam Ford, and Joshua Green, *Spreadable Media: Creating Value and Meaning in a Networked Culture* (New York: New York University Press, 2013).

62. "Get Discovered," YouTube, accessed May 23, 2016, https://storage.googleapis.com/creator-academy-assets/discovery/discovery%20-%20en.pdf.

63. "Reach Beyond YouTube," YouTube, accessed May 23, 2016, https://storage.googleapis.com/creator-academy-assets/reach-beyond/reach-beyond%20-%20en.pdf.

64. For further discussion of this point, see Ramon Lobato and Julian Thomas, *The Informal Media Economy* (Cambridge, UK: Polity Press, 2015).

65. Jenkins, Ford, and Green, *Spreadable Media*.

66. Alex Needham, "Richard Prince v Suicide Girls in an Instagram Price War," *The Guardian*, May 27, 2015, https://www.theguardian.com/artanddesign/2015/may/27/suicide-girls-richard-prince-copying-instagram.

67. Jessica Contrera, "A Reminder that Your Instagram Photos Aren't Really Yours: Someone Else Can Sell Them for $90,000," *Washington Post*, May 25, 2015, https://www.washingtonpost.com/news/arts-and-entertainment/wp/2015/05/25/a-reminder-that-your-instagram-photos-arent-really-yours-someone-else-can-sell-them-for-90000/.

68. Ibid.

69. Matilda Battersby, "Stolen Instagram Photo Sells for $90,000," *The Independent*, May 25, 2015, http://www.independent.co.uk/arts-entertainment/art/news/stolen-instagram-photo-sells-for-90000-10274267.html.

70. Allen Murabayashi, "Opinion: Richard Prince Is a Jerk," PetaPixel, May 26, 2015, http://petapixel.com/2015/05/26/richard-prince-is-a-jerk/.

71. Sarah Thomas, "Australian Photographer Peter Coulson 'Violated' by Richard Prince Instagram Art Scandal," *Sydney Morning Herald*, May 27, 2015, http://www.smh.com.au/entertainment/art-and-design/australian-photographer-peter-coulson-violated-by-richard-prince-instagram-art-scandal-20150528-ghatkt.html.

72. Madison Malone Kircher, "It Turns Out Two of the Women Whose Instagram Photos Were Hijacked by Richard Prince Didn't Even Shoot the Originals," *Business Insider Australia*, May 29, 2015, http://www.businessinsider.com.au/instagrammers-from-richard-princes-photos-didnt-take-the-original-pictures-2015-5.

73. Cait Munro, "Richard Prince Instagram Victims Speak Out," ArtNetNews,May 29, 2015, https://news.artnet.com/art-world/more-richard-prince-instagram-303166.

74. Ibid.

75. Selena Mooney, Instagram post, accessed May 23, 2016, https://www.instagram .com/p/3MsKqdHWX1/.

76. Ibid.

77. Richard Prince, Twitter post, May 28, 2015, https://twitter.com/RichardPrince4 /status/603874714201751552?ref_src=twsrc%5Etfw.

78. See Adler, "Against Moral Rights," *California Law Review* 91, no. 1 (2009): 263–300.

79. Brian Larkin, *Signal and Noise: Media, Infrastructure, and Urban Culture in Nigeria* (Durham, NC: Duke University Press, 2008), 219. See also Lars Eckstein and Anja Schwarz, *Postcolonial Piracy: Media Distribution and Cultural Production in the Global South* (London and New York: Bloomsbury Publishing, 2014); Lawrence Liang, "Cinematic Citizenship and the Illegal City," *Inter-Asia Cultural Studies* 3, no. 6 (2005): 366–385; Ramon Lobato, *Shadow Economies of Cinema: Mapping Informal Film Distribution* (London: Palgrave Macmillan, 2012); and Ravi Sundaram, *Pirate Modernity: Delhi's Media Urbanism* (New York: Routledge, 2010).

80. Joe Karaganis, "Rethinking Piracy" in *Media Piracy in Emerging Economies*, ed. Joe Karaganis (New York: Social Science Research Council, 2011), 43.

81. Sundaram, *Pirate Modernity*, 112.

82. See Larkin, *Signal and Noise*, 240.

83. Lawrence Liang and Achal Prabhala, "Reconsidering the Pirate Nation: Comment," *African Journal of Information and Communication* 7 (2006): 110.

84. Ibid., 111.

85. Lawrence Liang and Ravi Sundaram, "India," in *Media Piracy in Emerging Economies*, ed. Joe Karaganis (New York: Social Science Research Council, 2011), 344.

86. Lobato, *Shadow Economies of Cinema*.

87. Ibid., 89.

88. Ibid.

89. Ibid.

90. Ibid.

91. Justin Hughes, "The Philosophy of Intellectual Property," *Georgetown Law Journal* 77, no. 2 (1988): 287–366; and Peter Drahos, *A Philosophy of Intellectual Property*, (Aldershot, UK: Ashgate Publishing, 1996).

92. See Carys Craig, *Copyright, Communication and Culture: Towards a Relational Theory of Copyright Law* (Cheltenham, UK: Edward Elgar Publishing, 2011).

Chapter 6

1. See Rosemary Coombe, *The Cultural Life of Intellectual Properties: Authorship, Appropriation and the Law* (Durham, NC: Duke University Press, 1998).

2. See A&M Records, Inc. v. Napster, Inc. 239 F.3d 1004 (9th Cir. 2001); and A&M Records, Inc. v. Napster, Inc. 114 F. Supp. 2d 896 (2000).

3. Universal Music Pty Ltd. v. Sharman License Holdings [2005] FCA 1242.

4. Roadshow Films Pty Ltd. & Ors v. iiNet Ltd. [2010] FCA 24; Roadshow Films Pty Ltd. & Ors v. iiNet Ltd. [2011] FCAFC 24; and Roadshow Films Pty Ltd. & Ors v. iiNet Ltd. [2012] HCA 16.

5. Dallas Buyers Club LLC v. iiNet Ltd. [2015] FCA 317; Dallas Buyers Club LLC v. iiNet Ltd. (No 3) [2015] FCA 422; Dallas Buyers Club LLC v. iiNet Ltd. (No 4) [2015] FCA 838; and Dallas Buyers Club LLC v. iiNet Ltd. (No 5) [2015] FCA 1437 (16 December 2015).

6. See generally Bryan Choi, "The Grokster Dead-End," *Harvard Journal of Law & Technology* 19, no. 2 (2006): 393–411; Rebecca Giblin, *Code Wars: 10 Years of P2P Software Litigation* (Cheltenham, UK: Edward Elgar Publishing, 2011); and Matthew Rimmer, *Digital Copyright and the Consumer Revolution: Hands Off My iPod* (Cheltenham, UK: Edward Elgar Publishing, 2007).

7. Giblin, *Code Wars*, 15.

8. Ibid.

9. Stacey Lantagne, "The Morality of MP3s: The Failure of the Recording Industry's Plan of Attack," *Harvard Journal of Law & Technology* 18, no. 1 (2004): 272.

10. Most famously, Metallica went so far as to stage a public protest outside Napster's headquarters. See Lee Marshall, "Metallica and Morality: The Rhetorical Battleground of the Napster Wars," *Entertainment Law* 1, no. 1 (2002): 1–19.

11. A&M Records, Inc. v. Napster, Inc. 114 F. Supp. 2d 896 (2000) at 1.

12. Ibid.

13. The dangers of this interpretive approach to innovative technologies have been explored in Jessica Litman, *Digital Copyright* (New York: Prometheus Books, 2001); Rimmer, *Digital Copyright and the Consumer Revolution*; and Sarah Holthusen, "The Napster Decision: Implications for Copyright Law in the Digital Age," *University of Queensland Law Journal* 21, no. 2 (2001): 245–250.

14. Sony Corp. of America v. Universal City Studios Inc., 464 U.S. 417 104 S. Ct. 774, 78 L. Ed. 2d 574 (1984).

15. *Home Recording of Copyrighted Works: Hearings before the Subcommittee on Courts, Civil Liberties, and the Administration of Justice of the Committee on the Judiciary, House of Representatives, Ninety-seventh Congress, second session, on H.R. 4783, H.R. 4794 H.R. 4808, H.R. 5250, H.R. 5488, and H.R. 5705* (1982), last updated May 30, 2002, http://cryptome.org/hrcw-hear.htm.

16. Sony Corp. of America v. Universal City Studios, Inc., 464 U.S. 417, 104 S. Ct. 774, 78 L. Ed. 2d 574 (1984) at 52.

17. Kathy Bowrey and Matthew Rimmer, "Rip, Mix, Burn: The Politics of Peer to Peer and Copyright Law," *First Monday* 7, no. 8 (2002), available at firstmonday.org/article/view/974/895.

18. A&M Records, Inc. v. Napster, Inc. 239 F.3d 1004 (9th Cir. 2001) at 49.

19. See Holthusen, "The Napster Decision," for further details.

20. David Kravets, "Napster Trial Ends Seven Years Later, Defining Online Sharing along the Way," *Wired* accessed August 31, 2007, http://www.wired.com/2007/08/napster-trial-e/.

21. A&M Records, Inc. v. Napster, Inc. 114 F. Supp. 2d 896 (2000) at 2.

22. Ibid., at 61.

23. Ibid., at 42, 46.

24. Ibid., at 10.

25. Ibid., at 13.

26. Ibid., at 69.

27. Fonovisa Inc. v. Cherry Auction Inc. 76 F.3d. 259, 37 U.S.P.Q.2d 1590, 9th Cir. 1996.

28. Gershwin Publishing Corp. v. Columbia Artists Management 443 F. 2d 1159, 1162, 2d Cir. 1971.

29. This is not a novel claim. The best brief articulation of this argument is found in Lucas Hilderbrand, *Inherent Vice: Bootleg Histories of Videotape and Copyright* (Durham, NC: Duke University Press, 2009), 109–113.

30. A&M Records, Inc. v. Napster, Inc. 114 F. Supp. 2d 896 (2000) at 2.

31. John Logie, *Peers, Pirates, and Persuasion: Rhetoric in the Peer-to-Peer Debates* (West Lafayette, IN: Parlor Press, 2006), 58–59.

32. Kylie Pappalardo, "A Tort Law Framework for Copyright Authorisation" (PhD thesis, Australian Catholic University, 2015), 92.

33. A&M Records, Inc. v. Napster, Inc. 114 F. Supp. 2d 896 (2000) at 13.

34. Ibid., at 31, 35.

35. Ibid.

36. Choi, "The Grokster Dead-End," 395–396.

37. Universal Music Pty Ltd. v. Sharman License Holdings [2005] FCA 1242. This framing of Kazaa as an Australian service necessarily simplifies the complicated structuring of the company across its history. As Matthew Rimmer explains: "In January 2002, while related legal action was pending against it in the Netherlands, Kazaa BV transferred ownership of key assets to the newly formed Sharman Networks, Ltd. ... Sharman is a company organized under the laws of the island-nation of Vanuatu and doing business principally in Australia." See Matthew Rimmer, "Hail to the Thief: A Tribute to Kazaa," *University of Ottawa Law and Technology Journal* 2, no. 1 (2005): 180.

38. Universal Music Pty Ltd. v. Sharman License Holdings [2005] FCA 1242 at 54.

39. Ibid., at 151.

40. Ibid.

41. Ibid., at 267.

42. Ibid.

43. Ibid., at 405.

44. Ibid., at 406.

45. Ibid., at 420.

46. Giblin, *Code Wars*, 136.

47. Pappalardo, "A Tort Law Framework for Copyright Authorisation," 110.

48. Universal Music Pty Ltd. v. Sharman License Holdings [2005] FCA 1242 at 340.

49. Pappalardo, "A Tort Law Framework for Copyright Authorisation," 99.

50. Apple, *Apple Unveils New iMacs with CD-RW Drives & iTunes Software*, press release, February 22, 2001, https://www.apple.com/pr/library/2001/02/22Apple-Unveils-New-iMacs-With-CD-RW-Drives-iTunes-Software.html.

51. Jean Burgess, "The iPhone Moment, the Apple Brand and the Creative Consumer: From 'Hackability and Usability' to Cultural Generativity," in *Studying Mobile*

Media: Cultural Technologies, Mobile Communication, and the iPhone, ed. Larissa Hjorth, Ingrid Richardson, and Jean Burgess (New York: Routledge, 2012), 35.

52. Peter Cohen, "Disney Boss Accuses Apple of Fostering Piracy," Macworld, March 1, 2002, http://www.macworld.com/article/1003743/eisner.html.

53. Universal Music Pty Ltd. v. Sharman License Holdings [2005] FCA 1242 at 154.

54. Ibid.

55. See Catherine Bond, Abi Paramaguru, and Graham Greenleaf, "Advance Australia Fair? The Copyright Reform Process," *Journal of World Intellectual Property* 10, no. 3–4 (2007): 284–313; James Meese, "Resistance or Negotiation: An Australian Perspective on Copyright Law's Cultural Agenda," *Computers and Composition* 27, no. 3 (2010): 167–178; Matthew Rimmer, "Robbery under Arms: Copyright Law and the Australia-United States Free Trade Agreement," *First Monday* 11, no. 3 (2006), available at firstmonday.org/ojs/index.php/fm/article/view/1316; and Kimberlee Weatherall, "Of Copyright Bureaucracies and Incoherence: Stepping Back from Australia's Recent Copyright Reforms," *Melbourne University Law Review* 31, no. 3 (2007): 967–1174.

56. See Goldwyn-Mayer Studios Inc. v. Grokster, Ltd. 125 S. Ct. 2764 (2005) and In re Aimster Copyright Litigation 333 F.3d 643 (7th Cir. 2003).

57. Giblin, *Code Wars*, 98, and see chapter 5 of this book.

58. Giblin, *Code Wars*, 103.

59. See ibid., 104–126.

60. This reform process has been detailed by Bond, Paramaguru, and Greenleaf, "Advance Australia Fair?"; Meese, "Resistance or Negotiation"; Rimmer, "Robbery under Arms"; and Weatherall, "Of Copyright Bureaucracies and Incoherence."

61. Copyright Act 1968 (Cth), s39B.

62. Copyright Act 1968 (Cth), s116AH.

63. See Section 512 of the Online Copyright Infringement Liability Limitation Act or 17 U.S.C. §512.

64. Ibid.

65. Laurence Helfer, "Regime Shifting: the TRIPs Agreement and New Dynamics of International Intellectual Property Lawmaking," *Yale Journal of International Law* 29, no. 1 (2004): 7. Also see Blayne Haggart, *Copyfight: The Global Politics of Digital Copyright Reform* (Toronto: University of Toronto Press, 2014).

66. Clive Thompson, "The BitTorrent Effect," *Wired*, January 1, 2005, http://www.wired.com/2005/01/bittorrent-2/.

67. David Lindsay, "Liability of ISPs for End-User Copyright Infringements: The High Court Decision in Roadshow Films v. iiNet," *Telecommunications Journal of Australia* 62, no. 4 (2012): 53.1.

68. Robert McCallum Jr., "Film/TV Industry Files Copyright Case against Aussie ISP," Diplomatic cable, November 30, 2008, reference ID: 08CANBERRA1197, accessed July 23, 2013, http://cablegatesearch.net/cable.php?id=08CANBERRA1197&q=iinet.

69. "Our Story," Motion Picture Association of America, accessed May 21, 2016, http://www.mpaa.org/our-story/.

70. McCallum Jr., "Film/TV Industry Files Copyright Case against Aussie ISP."

71. Ibid.

72. Lindsay, "Liability of ISPs for End-User Copyright Infringements," 53.8.

73. Roadshow Films Pty Ltd. & Ors v. iiNet Ltd. (2010) FCA 24 at 97.

74. Ibid.

75. Ibid., at 99.

76. Ibid., at 179.

77. Ibid., at 401.

78. Ibid., at 411.

79. Roadshow Films Pty Ltd. & Ors v. iiNet Ltd. [2011] HCA 16 at 77.

80. Ibid., at 78.

81. Ibid.

82. Dallas Buyers Club LLC v iiNet Ltd. [2015] FCA 317 at 1.

83. Ibid., at 2.

84. Ibid., at 3.

85. For a critical view of this practice, see TorrentFreak, *The Speculative Invoicing Handbook* (Self-Published, 2009).

86. Ibid., at 86–87.

87. Ibid., at 87.

88. Dallas Buyers Club LLC v. iiNet Ltd. (No 3) [2015] FCA 422 at 6.

89. Dallas Buyers Club LLC v. iiNet Ltd. (No 4) [2015] FCA 838 at 13.

90. Ibid., at 15.

91. Ibid., at 15, 23.

92. Ibid., at 32.

93. Ibid., at 35.

94. Dallas Buyers Club LLC v. iiNet Ltd. (No 5) [2015] FCA 1437 at 3, 4, 10.

95. Ibid., at 47.

96. Ibid., at 52.

97. Ibid., at 55.

98. James Meese, "The Pirate Imaginary and the Potential of the Authorial Pirate," in *Piracy: Leakages from Modernity*, ed. Martin Fredriksson and James Avanitakis (Sacramento, CA: Litwin Books, 2014), 19–37.

99. Dallas Buyers Club LLC v. iiNet Ltd. (No 3) [2015] FCA 422 at 422.

100. Mark Deuze, *Media Life* (Cambridge, UK: Polity, 2012).

101. See Raphaël Nowak, *Consuming Music in the Digital Age: Technologies, Roles and Everyday Life* (London: Palgrave Macmillan, 2016), 20–28.

102. Mike Vago, "Louis CK Releases Pay-What-You-Want Live Album," A. V. Club, August 11, 2015, http://www.avclub.com/article/louis-ck-releases-pay-what-you-want-live-album-223638.

103. Helienne Lindvall, "How Record Labels Are Learning to Make Money from YouTube," *The Guardian*, January 4, 2013, http://www.theguardian.com/media/2013/jan/04/record-labels-making-money-youtube.

104. Amy X. Wang, "Recorded Music Is in Trouble, but Live Concerts Are Making More Money Than Ever," Quartz, February 26, 2016, http://qz.com/625934/recorded-music-is-in-trouble-but-live-concerts-are-making-more-money-than-ever/. Of course, alongside these changes fans are also increasingly fetishizing physical cultural artifacts. See Dominik Bartmanski and Ian Woodward, "The Vinyl: The Analogue Medium in the Age of Digital Reproduction," *Journal of Consumer Culture* 15, no. 1 (2015): 3–27.

105. Stuart Dredge, "How Much Do Musicians Really Make from Spotify, iTunes and YouTube?," *The Guardian*, April 3, 2015, http://www.theguardian.com/technology/2015/apr/03/how-much-musicians-make-spotify-itunes-youtube.

106. See generally Ramon Lobato and James Meese, eds., *Geoblocking and Global Video Culture* (Amsterdam: Institute of Network Cultures, 2016).

107. Anti-Counterfeiting Trade Agreement, at E-1, Oct. 1, 2011, 50 I.L.M. 239, 243 (2011) art. 5(k).

108. ACTA, at art. 23, § 1.

109. Michael Carrier, "SOPA, PIPA, ACTA, TPP: An Alphabet Soup of Innovation-Stifling Copyright Legislation and Agreements," *Northwestern Journal of Technology and Intellectual Property* 11, no. 2 (2013): 21–31. See also Matthew Rimmer, "Trick or Treaty?: The Australian Debate over the Anti-Counterfeiting Trade Agreement," in *The ACTA and the Plurilateral Enforcement Agenda: Genesis and Aftermath*, ed. Pedro Roffe and Xavier Seuba (Cambridge, UK: Cambridge University Press, 2015), 169–201.

110. See Lobato and Meese, *Geoblocking and Global Video Culture*.

111. Angela Daly, *Socio-Legal Aspects of the 3D Printing Revolution* (London: Palgrave Pivot, 2016); Luke Heemsbergen et al., "Social Practices of 3D Printing: Decentralising Control and Reconfiguring Regulation," *Australian Journal of Telecommunications and the Digital Economy* 4, no. 3 (2016): 110–125; and Adam Rugg and Ben Burroughs, "Periscope, Live Streaming and Mobile Video Culture," in *Geoblocking and Global Video Culture*, ed. Ramon Lobato and James Meese (Amsterdam: Institute of Network Cultures, 2016), 64–73.

Conclusion

1. Carys Craig, *Copyright, Communication and Culture: Towards a Relational Theory of Copyright Law* (Cheltenham, UK: Edward Elgar Publishing, 2011), 244.

2. See William Patry, *Moral Panics and the Copyright Wars* (Oxford, UK: Oxford University Press, 2009).

3. Ibid., 103.

4. See Laura J. Murray, S. Tina Piper, and Kirsty Robertson, *Putting Intellectual Property in Its Place: Rights Discourses, Creative Labor, and the Everyday* (Oxford, UK: Oxford University Press, 2014).

5. See generally Jennifer Urban, Joe Karaganis, and Brianna Schofield, *Notice and Takedown in Everyday Practice* (Berkley, CA: Berkley Law, 2016).

6. Tarleton Gillespie, *Wired Shut: Copyright and the Shape of Digital Culture* (Cambridge, MA: MIT Press, 2007). See also Julie Cohen, "Pervasively Distributed Copyright Enforcement," *Georgetown Law Journal* 95, no. 1 (2006): 1–48.

7. James Meese et al., "Entering the Graveyard Shift: Disassembling the Australian TiVo," *Television & New Media* 16, no. 2 (2015): 167.

8. Ibid.

9. Ian Hargreaves, *Digital Opportunity: A Review of Intellectual Property and Growth* (London: Department for Business, Innovation and Skills, 2011), 17.

10. Ruth Towse, "The Quest for Evidence on the Economic Effects of Copyright Law," *Cambridge Journal of Economics* 37, no. 5 (2013): 1187–1202. See also Patricia

Aufderheide and Dorian Hunter Davis, "Contributors and Arguments in Australian Policy Debates on Fair Use and Copyright: The Missing Discussion of the Creative Process," *International Journal of Communication* 11 (2017): 522–545.

11. Kathy Bowrey has advanced this argument in "Law, Aesthetics and Copyright Historiography: A Critical Reading of the Genealogies of Martha Woodmansee and Mark Rose," in *Research Handbook on the History of Copyright Law*, ed. Isabella Alexander and Tomás Gómez-Arostegui (Cheltenham, UK: Edward Elgar Publishing, 2016), 27–52; also see Julie Cohen, "The Place of the User in Copyright Law," *Fordham Law Review* 74 (2005): 347–374.

Bibliography

Adler, Amy. "Against Moral Rights." *California Law Review* 91 (1) (2009): 263–300.

Adler, Amy. "Fair Use and the Future of Art." *New York University Law Review* 91 (3) (2016): 559–626.

Akrich, Madeline. "The De-Scription of Technical Objects." In *Shaping Technology, Building Society: Studies in Sociotechnical Change*, edited by Wiebe Bijker and John Law, 205–224. Cambridge, MA: MIT Press, 1992.

Alexander, Isabella. "All Change for the Digital Economy: Copyright and Business Models in the Early Eighteenth Century." *Berkeley Technology Law Journal* 25 (3) (2010): 1351–1380.

Alexander, Isabella. *Copyright Law and the Public Interest in the Nineteenth Century.* Oxford, UK and Portland, OR: Hart Publishing, 2010.

Althusser, Louis. *Lenin and Philosophy and Other Essays.* New York: Monthly Review Press, 1971.

Andersson Schwarz, Jonas, and Patrick Burkart. "Piracy and Social Change—Revisiting Pirate Cultures." *International Journal of Communication* 9 (2015): 792–797.

Ang, Ien. *Desperately Seeking the Audience.* London: Routledge, 1991.

Ang, Ien. *Living Room Wars: Rethinking Media Audiences for a Postmodern World.* London: Routledge, 1996.

Arewa, Olufunmilayo. "From JC Bach to Hip Hop: Musical Borrowing, Copyright and Cultural Context." *North Carolina Law Review* 84 (2) (2006): 547–645.

Aufderheide, Patricia, and Dorian Hunter Davis. "Contributors and Arguments in Australian Policy Debates on Fair Use and Copyright: The Missing Discussion of the Creative Process." *International Journal of Communication* 11 (2017): 522–545.

Aufderheide, Patricia, and Peter Jaszi. *Reclaiming Fair Use: How to Put Balance Back in Copyright.* Chicago: University of Chicago Press, 2011.

Bakardjieva, Maria. *Internet Society: The Internet in Everyday Life*. London: Sage Publications, 2005.

Barthes, Roland. *Image-Music-Text*. London: Fontanta, 1977.

Bartmanski, Dominik, and Ian Woodward. "The Vinyl: The Analogue Medium in the Age of Digital Reproduction." *Journal of Consumer Culture* 15 (1) (2015): 3–27.

Benkler, Yochai. *The Wealth of Networks: How Social Production Transforms Markets and Freedom*. New Haven, CT: Yale University Press, 2006.

Bently, Lionel, and Laura Biron. "Discontinuities between Legal Conceptions of Authorship and Social Practices." In *The Work of Authorship*, edited by Mireille van Eechoud, 237–276. Amsterdam: University of Amsterdam Press, 2014.

Blackman, Lisa, John Cromby, Derek Hook, Dimitris Papadopoulos, and Valerie Walkerdine. "Creating Subjectivities." *Subjectivity* 22 (1) (2008): 1–27.

Bond, Catherine, Abi Paramaguru, and Graham Greenleaf. "Advance Australia Fair? The Copyright Reform Process." *Journal of World Intellectual Property* 10 (3–4) (2007): 284–313.

Boon, Marcus. *In Praise of Copying*. Cambridge, MA: Harvard University Press, 2011.

Borschke, Margie. *This Is Not a Remix: Piracy, Authenticity and Popular Music*. London: Bloomsbury Publishing, 2017.

Bowrey, Kathy. "Copyright, the Paternity of Artistic Works, and the Challenge Posed by Postmodern Artists." *Intellectual Property Journal* 8 (3) (1994): 285–317.

Bowrey, Kathy. "Law, Aesthetics and Copyright Historiography: A Critical Reading of the Genealogies of Martha Woodmansee and Mark Rose." In *Research Handbook on the History of Copyright Law*, edited by Isabella Alexander and Tomás Gómez-Arostegui, 27–52. Cheltenham, UK: Edward Elgar Publishing, 2016.

Bowrey, Kathy. "On Clarifying the Role of Originality and Fair Use in 19th Century UK Jurisprudence: Appreciating 'The Humble Grey which Emerges as the Result of Long Controversy.'" In *The Common Law of Intellectual Property: Essays in Honour of Professor David Vaver*, edited by Lionel Bently, Catherine Ng, and Giuseppina D'Agostino, 45–72. Oxford, UK: Hart Publishing, 2010.

Bowrey, Kathy. "Who's Writing Copyright's History?" *European Intellectual Property Review* 18 (6) (1996): 322–329.

Bowrey, Kathy, and Jane Anderson. "The Politics of global Information Sharing: Whose Cultural Agendas Are Being Advanced?" *Social & Legal Studies* 18 (4) (2009): 479–504.

Bowrey, Kathy, and Natalie Fowell. "Digging Up Fragments and Building IP Franchises." *Sydney Law Review* 31 (2) (2009): 185–210.

Bowrey, Kathy, and Michael Handler. "Instituting Copyright: Reconciling Copyright Law and Industry Practice in the Australian Film and Television Sector." In *Law and Creativity in the Age of the Entertainment Franchise*, edited by Kathy Bowrey and Michael Handler, 140–169. Cambridge, UK: Cambridge University Press, 2014.

Bowrey, Kathy, and Matthew Rimmer. "Rip, Mix, Burn: The Politics of Peer to Peer and Copyright Law." *First Monday* 7 (8) (2002). Available at firstmonday.org/article /view/974/895.

Boyle, James. "The Second Enclosure Movement and the Construction of the Public Domain." *Law and Contemporary Problems* 66 (1/2) (2003): 33–74.

Bracha, Oren. "The Ideology of Authorship Revisited: Authors, Markets, and Liberal Values in Early American Copyright." *Yale Law Journal* 118 (2) (2008): 186–271.

Bruns, Axel. *Blogs, Wikipedia, Second Life, and Beyond: From Production to Produsage.* New York: Peter Lang, 2008.

Buccafusco, Christopher. "A Theory of Copyright Authorship." *Virginia Law Review* 102 (5) (2016): 1229–1296.

Burgess, Jean. "From 'Broadcast Yourself' to 'Follow Your Interests': Making Over Social Media." *International Journal of Cultural Studies* 18 (3) (2015): 281–285.

Burgess, Jean. "The iPhone Moment, the Apple Brand and the Creative Consumer: From 'Hackability and Usability' to Cultural Generativity." In *Studying Mobile Media: Cultural Technologies, Mobile Communication, and the iPhone*, edited by Larissa Hjorth, Ingrid Richardson, and Jean Burgess, 28–42. New York: Routledge, 2012.

Burgess, Jean, and Joshua Green. *YouTube: Online Video and Participatory Culture.* Cambridge, UK: Polity Press, 2012.

Burkart, Patrick. *Pirate Politics.* Cambridge, MA: The MIT Press, 2014.

Carrier, Michael. "SOPA, PIPA, ACTA, TPP: An Alphabet Soup of Innovation-Stifling Copyright Legislation and Agreements." *Northwestern Journal of Technology and Intellectual Property* 11 (2) (2013): 21–31.

Chander, Anupam, and Madhavi Sunder. "Copyright's Cultural Turn." *Texas Law Review* 91 (2013): 1397–1412.

Chander, Anupam, and Madhavi Sunder. "Everyone's a Superhero: A Cultural Theory of 'Mary Sue' Fan Fiction as Fair Use." *California Law Review* 95 (2) (2007): 597–626.

Choi, Bryan. "The Grokster Dead-End." *Harvard Journal of Law & Technology* 19 (2) (2006): 393–411.

Cohen, Julie. *Configuring the Networked Self: Law, Code, and the Play of Everyday Practice.* New Haven, CT: Yale University Press, 2012.

Cohen, Julie. "Creativity and Culture in Copyright Theory." *U.C. Davis Law Review* 40 (3) (2007): 1151–1205.

Cohen, Julie. "Pervasively Distributed Copyright Enforcement." *Georgetown Law Journal* 95 (1) (2006): 1–48.

Cohen, Julie. "The Place of the User in Copyright Law." *Fordham Law Review* 74 (2005): 347–374.

Commonwealth Attorney General, and Phillip Ruddock. *Major Copyright Reforms Strike Balance*, press release, 2006.

Coombe, Rosemary. *The Cultural Life of Intellectual Properties: Authorship, Appropriation and the Law*. Durham, NC: Duke University Press, 1998.

Coombe, Rosemary. "Room for Manoeuver: Toward a Theory of Practice in Critical Legal Studies." *Law & Social Inquiry* 14 (1) (1989): 69–121.

The Copyright Law Review Committee. *Report of the Committee Appointed by the Attorney General of the Commonwealth to Consider What Alterations Are Desirable in the Copyright Law of the Commonwealth*. Australia, 1958.

Craig, Carys. *Copyright, Communication and Culture: Towards a Relational Theory of Copyright Law*. Cheltenham, UK: Edward Elgar Publishing, 2011.

Craig, Carys. "Reconstructing the Author-Self: Some Feminist Lessons for Copyright Law." *Journal of Gender, Social Policy & the Law* 15 (2) (2007): 207–268.

Cubitt, Sean. "Distribution and Media Flows." *Cultural Politics* 1 (2) (2005): 193–214.

Currah, Andrew. "Hollywood, the Internet and the World: A Geography of Disruptive Innovation." *Industry and Innovation* 14 (4) (2007): 359–384.

Daly, Angela. *Socio-Legal Aspects of the 3D Printing Revolution*. London: Palgrave Pivot, 2016.

Darling, Kate. "IP without IP: A Study of the Online Adult Entertainment Industry." *Stanford Technology Law Review* 17 (3) (2013): 709–771.

Deazley, Ronan. *On the Origin of the Right to Copy: Charting the Movement of Copyright Law in Eighteenth Century Britain (1695–1775)*. Oxford, UK: Hart Publishing, 2004.

Deazley, Ronan. "The Statute of Anne and the Great Abridgement Swindle." *Houston Law Review* 47 (4) (2010): 793–818.

Defoe, Daniel. *An Essay on the Regulation of the Press*. London, 1704.

DelNero, Matthew. "Long Overdue? An Exploration of the Status and Merit of a General Public Performance Right in Sound Recordings." *Journal of the Copyright Society of the U.S.A.* 51 (3) (2004): 473–520.

Deuze, Mark. *Media Life*. Cambridge, UK: Polity, 2012.

Deveci, Hasan. "Databases: Is Sui Generis a Stronger Bet Than Copyright?" *International Journal of Law and Information Technology* 12 (2) (2004): 178–208.

Drahos, Peter. "BITs and BIPs: Bilateralism in Intellectual Property." *Journal of World Intellectual Property* 4 (6) (2001): 791–808.

Drahos, Peter. *A Philosophy of Intellectual Property*. Aldershot, UK: Ashgate Publishing, 1996.

Drassinower, Abraham. "From Distribution to Dialogue: Remarks on the Concept of Balance in Copyright Law." *Journal of Corporation Law* 34 (4) (2009): 991–1007.

Drassinower, Abraham. "A Rights-Based View of the Idea/Expression Dichotomy in Copyright Law." *Canadian Journal of Law and Jurisprudence* 16 (1) (2003): 3–21.

Durham, Alan. "The Random Muse: Authorship and Indeterminacy." *William and Mary Law Review* 44 (2) (2002): 569–642.

Eckstein, Lars, and Anja Schwarz. *Postcolonial Piracy: Media Distribution and Cultural Production in the Global South*. London and New York: Bloomsbury Publishing, 2014.

Edmond, Maura. "Here We Go Again: Music Videos after YouTube." *Television & New Media* 15 (4) (2014): 305–320.

Edwards, Lee, Bethany Klein, David Lee, Giles Moss, and Fiona Philip. "Framing the Consumer: Copyright Regulation and the Public." *Convergence: The International Journal of Research into New Media Technologies* 19 (1) (2013): 9–24.

Engelman, Ralph. *Public Radio and Television in America: A Political History*. California: SAGE Publications, 1996.

Erickson, Kristofer. "User Illusion: Ideological Construction of 'User-Generated Content' in the EC Consultation on Copyright." *Internet Policy Review* 3 (4) (2014): 1–19.

Fauchart, Emmanuelle, and Eric Von Hippel. "Norms-Based Intellectual Property Systems: The Case of French Chefs." *Organization Science* 19 (2) (2008): 187–201.

Feather, John. *A History of British Publishing*. London: Routledge, 2006.

Feather, John. *Publishing, Piracy, and Politics: A Historical Study of Copyright in Britain*. New York: Mansell, 1994.

Fisher, William, III. "Reconstructing the Fair Use Doctrine." *Harvard Law Review* 101 (8) (1988): 1659–1795.

Fisk, Catherine. "Authors at Work: The Origins of the Work-for-Hire Doctrine." *Yale Journal of Law & the Humanities* 15 (1) (2013): 1–70.

Fisk, Catherine. "Will Work for Screen Credit: Labour and the Law in Hollywood." In *Hollywood and the Law*, edited by Paul McDonald, Emily Carman, Eric Hoyt, and Philip Drake, 235–262. London: Palgrave, 2015.

Foucault, Michel. "What Is an Author." In *Aesthetics, Methods and Epistemology*, edited by James D. Faubion, 205–222. New York: The New Press, 1998.

Fredriksson, Martin. "Copyright Culture and Pirate Politics." *Cultural Studies* 28 (5–6) (2014): 1022–1047.

Fredriksson, Martin. "Piracy, Globalisation and the Colonisation of the Commons." *Global Media Journal: Australian Edition* 6 (1) (2012). Available at http://www.hca.westernsydney.edu.au/sites/wp_gmjau/archive/v6_2012_1/pdf/martin_fredriksson_RA_6_1_2012.pdf

Fredriksson, Martin. "The Pirate Party and the Politics of Communication." *International Journal of Communication* 9 (2015): 909–924.

Geiger, Christophe. "From Berne to National Law, via the Copyright Directive: The Dangerous Mutations of the Three-Step Test." *European Intellectual Property Review* 29 (12) (2007): 486–491.

Geiger, Christophe, Daniel Gervais, and Martin Senftleben. "The Three-Step Test Revisited: How to Use the Test's Flexibility in National Copyright Law." *American University International Law Review* 29 (3) (2014): 581–626.

Geist, Michael, ed. *The Copyright Pentalogy*. Ottawa: University of Ottawa Press, 2013.

Geist, Rachel Ann. "A License to Read: The Effect of E-Books on Publishers, Libraries, and the First Sale Doctrine." *Idea* 52 (1) (2012): 63–100.

Gelman, Eric, Janet Huck, Connie Leslie, Carolyn Friday, Pamela Abramson, and Michael Reese. "The Video Revolution." *Newsweek*. August 6, 1984.

Giblin, Rebecca. *Code Wars: 10 Years of P2P Software Litigation*. Cheltenham, UK: Edward Elgar Publishing, 2011.

Giblin, Rebecca. "Optus v NRL: A Seismic Shift for Time Shifting in Australia." *European Intellectual Property Review* 34 (5) (2012): 357–363.

Giblin, Rebecca, and Jane Ginsburg. "We (Still) Need to Talk about Aereo: New Controversies and Unresolved Questions after the Supreme Court's Decision." *Columbia Journal of Law & the Arts* 38 (2) (2015): 109–156.

Gibson, Johanna. *Creating Selves: Intellectual Property and the Narration of Culture*. Aldershot, UK: Ashgate Publishing, 2006.

Gillespie, Tarleton. "Characterizing Copyright in the Classroom: The Cultural Work of Antipiracy Campaigns." *Communication, Culture & Critique* 2 (3) (2009): 274–318.

Gillespie, Tarleton. *Wired Shut: Copyright and the Shape of Digital Culture*. Cambridge, MA: MIT Press, 2007.

Ginsburg, Jane. "The Concept of Authorship in Comparative Copyright Law." *DePaul Law Review* 52 (4) (2003): 1063–1092.

Ginsburg, Jane. "Tale of Two Copyrights: Literary Property in Revolutionary France and America." *Tulane Law Review* 64 (5) (1990): 991–1032.

Ginsburg, Jane. "'Une Chose Publique'? The Authors Domain and the Public Domain in Early British, French and U.S. Copyright Law." *Cambridge Law Journal* 65 (3) (2006): 636–670.

Gomez-Arostegui, H. Tomas. "Copyright at Common Law in 1774." *Connecticut Law Review* 47 (1) (2014): 1–57.

Gordon, Wendy. "Fair Use as Market Failure: A Structural and Economic Analysis of the 'Betamax' Case and Its Predecessors." *Columbia Law Review* 82 (8) (1982): 1600–1657.

Griffen-Foley, Bridget. "From Tit-Bits to Big Brother: A Century of Audience Participation in the Media." *Media, Culture & Society* 26 (4) (2004): 533–548.

Haggart, Blayne. *Copyfight: The Global Politics of Digital Copyright Reform*. Toronto: University of Toronto Press, 2014.

Halbert, Deborah. *Resisting Intellectual Property*. London: Routledge, 2006.

Hall, Donald. *Subjectivity*. London: Routledge, 2004.

Hamilton, James. "Historical Forms of User Production." *Media, Culture & Society* 36 (4) (2014): 491–507.

Hansmann, Henry, and Marina Santilli. "Authors' and Artists' Moral Rights: A Comparative Legal and Economic Analysis." *Journal of Legal Studies* 26 (1) (1997): 95–143.

Hargreaves, Ian. *Digital Opportunity: A Review of Intellectual Property and Growth*. London: Department for Business, Innovation and Skills, 2011.

Heemsbergen, Luke, Robbie Fordyce, Bjorn Nansen, Thomas Apperley, Mike Arnold, and Thomas Birtchnell. "Social Practices of 3D Printing: Decentralising Control and Reconfiguring Regulation." *Australian Journal of Telecommunications and the Digital Economy* 4 (3) (2016): 110–125.

Helfer, Laurence. "Regime Shifting: The TRIPs Agreement and New Dynamics of International Intellectual Property Lawmaking." *Yale Journal of International Law* 29 (1) (2004): 1–83.

Herman, Bill. *The Fight over Digital Rights: The Politics of Copyright and Technology*. Cambridge, UK: Cambridge University Press, 2013.

Heymann, Laura. "Everything Is Transformative: Fair Use and Reader Response." *Columbia Journal of Law and the Arts* 31 (4) (2008): 445–466.

Highfield, Tim. *Social Media and Everyday Politics*. Malden, MA: Polity, 2016.

Hilderbrand, Lucas. *Inherent Vice: Bootleg Histories of Videotape and Copyright*. Durham, NC: Duke University Press, 2009.

Hirtle, Peter, Emily Hudson, and Andrew Kenyon. *Copyright and Cultural Institutions: Guidelines for Digitizations for U.S. Libraries, Archives and Museums*. Ithaca: Cornell University Library Press, 2009.

Holthusen, Sarah. "The Napster Decision: Implications for Copyright Law in the Digital Age." *University of Queensland Law Journal* 21 (2) (2001): 245–250.

Hudson, Emily, and Andrew Kenyon. "Digital Access: The Impact of Copyright on Digitisation Practices in Australian Museums, Galleries, Libraries, and Archives." *University of New South Wales Law Journal* 30 (1) (2007): 12–52.

Hughes, Justin. "The Philosophy of Intellectual Property." *Georgetown Law Journal* 77 (2) (1988): 287–366.

Hughes, Robert. *The Shock of the New*. New York: Knopf, 1981.

Hunter, D., R. Lobato, M. Richardson, and J. Thomas, eds. *Amateur Media: Social, Cultural and Legal Perspectives*. Oxford, UK: Routledge, 2013.

Jaszi, Peter. "Toward a Theory of Copyright: The Metamorphoses of 'Authorship.'" *Duke Law Journal* 40 (2) (1991): 455–502.

"Jazz Has Got Copyright Law and That Ain't Good." *Harvard Law Review* 118 (6) (2005): 1940–1961.

Jenkins, Henry. *Convergence Culture: Where Old and New Media Collide*. New York: New York University Press, 2006.

Jenkins, Henry. *Fans, Bloggers, and Gamers: Exploring Participatory Culture*. New York: New York University Press, 2006.

Jenkins, Henry. "Rethinking 'Rethinking Convergence/Culture.'" *Cultural Studies* 28 (2) (2014): 267–297.

Jenkins, Henry, Sam Ford, and Joshua Green. *Spreadable Media: Creating Value and Meaning in a Networked Culture*. New York: New York University Press, 2013.

John, Nicholas. "File Sharing and the History of Computing: or, Why File Sharing Is Called 'File Sharing.'" *Critical Studies in Media Communication* 31 (3) (2014): 198–211.

Johns, Adrian. *Piracy: The Intellectual Property Wars from Gutenberg to Gates*. Chicago: University of Chicago Press, 2009.

Kaplan, Benjamin. *An Unhurried View of Copyright*. New York: Columbia University Press, 1967.

Karaganis, Joe. "Rethinking Piracy." in *Media Piracy in Emerging Economies*, edited by Joe Karaganis, 1–74. New York: Social Science Research Council, 2011.

Kennedy, Jenny. "Conceptual Boundaries of Sharing." *Information, Communication & Society* 19 (4) (2016): 461–474.

Kennedy, Jenny. "Rhetorics of Sharing: Data, Imagination and Desire." In *Unlike Us Reader: Social Media Monopolies and Their Alternatives*, edited by Geert Lovink and Miriam Rasch, 127–136. Amsterdam: Institute of Network Cultures, 2013.

Kim, Jin. "The Institutionalization of YouTube: From User-Generated Content to Professionally Generated Content." *Media, Culture & Society* 34 (1) (2012): 53–67.

Kirkpatrick, Graeme. *Critical Technology: A Social Theory of Personal Computing*. Aldershot, UK: Ashgate, 2004.

Klenz, William. "Brahms Op. 38; Piracy, Pillage, Plagiarism or Parody?" *Music Review* 34 (1) (1973): 39–50.

Koelman, Kamiel. "Fixing the Three-Step Test." *European Intellectual Property Review* 28 (8) (2006): 407–412.

Lantagne, Stacey. "The Morality of MP3s: The Failure of the Recording Industry's Plan of Attack." *Harvard Journal of Law & Technology* 18 (1) (2004): 269–293.

Larkin, Brian. "Degraded Images, Distorted Sounds: Nigerian Video and the Infrastructure of Piracy." *Public Culture* 16 (2) (2004): 289–314.

Larkin, Brian. *Signal and Noise: Media, Infrastructure, and Urban Culture in Nigeria*. Durham, NC: Duke University Press, 2008.

Latour, Bruno. "Where Are the Missing Masses? Sociology of a Few Mundane Artifacts." In *Shaping Technology, Building Society: Studies in Sociotechnical Change*, edited by Wiebe Bijker and John Law. 225–258. Cambridge, MA: MIT Press, 1992.

Leckey, Robert. *Contextual Subjects: Family, State and Relational Theory*. Toronto: University of Toronto Press, 2008.

Lee, Edward. *The Fight for the Future: How People Defeated Hollywood and Saved the Internet—For Now*. Self-Published, 2013. http://thefightforthefuture.com/wp-content/uploads/2013/07/The-Fight-for-the-Future_-How-P-Lee-Edward.pdf

Lessig, Lawrence. *Free Culture: How Big Media Uses Technology and the Law to Lock Down Culture and Control Creativity*. New York: Penguin Press, 2004.

Lessig, Lawrence. *Remix: Making Art and Commerce Thrive in the Hybrid Economy*. New York: Penguin, 2008.

Lessig, Lawrence. *Republic, Lost: How Money Corrupts Congress—and a Plan to Stop It.* New York: Twelve, 2011.

Lester, Toni. "Blurred Lines—Where Copyright Ends and Cultural Appropriation Begins—The Case of Robin Thicke versus Bridgeport Music and the Estate of Marvin Gaye." *Hastings Communications and Entertainment Law Journal* 36 (2) (2013): 217–242.

Leval, Pierre. "Nimmer Lecture: Fair Use Rescued." *UCLA Law Review* 44 (5) (June 1997): 1449–1467.

Liang, Lawrence. "Beyond Representation: The Figure of the Pirate." In *Access to Knowledge in the Age of Intellectual Property*, edited by Gaëlle Krikorian and Amy Kapczynski. 353–376. Cambridge, MA: MIT Press, 2010.

Liang, Lawrence. "Cinematic Citizenship and the Illegal City." *Inter-Asia Cultural Studies* 3 (6) (2005): 366–385.

Liang, Lawrence, and Achal Prabhala. "Reconsidering the Pirate Nation: Comment." *African Journal of Information and Communication* 7 (2006): 108–114.

Liang, Lawrence, and Ravi Sundaram. "India." In *Media Piracy in Emerging Economies*, edited by Joe Karaganis, 339–398. New York: Social Science Research Council, 2011.

Lindsay, David. "Liability of ISPs for End-User Copyright Infringements: The High Court Decision in Roadshow Films v. iiNet." *Telecommunications Journal of Australia* 62 (4) (2012): 53.1–53.24.

Litman, Jessica. *Digital Copyright.* New York: Prometheus Books, 2001.

Litman, Jessica. "The Politics of Intellectual Property." *Cardozo Arts and Entertainment Law Journal* 27 (2006): 313–320.

Litman, Jessica. "The Public Domain." *Emory Law Journal* 39 (4) (1990): 965–1023.

Litman, Jessica. "Sharing and Stealing." *Hastings Communications and Entertainment Law Journal* 27 (1) (2004): 1–48.

Liu, Joseph. "Copyright Law's Theory of the Consumer." *Boston College Law Review* 44 (2) (2003): 397–431.

Lobato, Ramon. "Constructing the Pirate Audience: On Popular Copyright Critique, Free Culture and Cyber-Libertarianism." *Media International Australia* 139 (1) (2011): 113–123.

Lobato, Ramon. "The Cultural Logic of Digital Intermediaries: YouTube Multichannel Networks." *Convergence: The International Journal of Research into New Media Technologies* 22 (4) (2016): 348–360.

Lobato, Ramon. *Shadow Economies of Cinema: Mapping Informal Film Distribution.* London: Palgrave Macmillan, 2012.

Lobato, Ramon, and James Meese, eds. *Geoblocking and Global Video Culture.* Amsterdam: Institute of Network Cultures, 2016.

Lobato, Ramon, and Julian Thomas. *The Informal Media Economy.* Cambridge, UK: Polity Press, 2015.

Lobato, Ramon, Julian Thomas, and Dan Hunter. "Histories of User-Generated Content: Between Formal and Informal Media Economies." *International Journal of Communication* 5 (2011): 899–914.

Logie, John. *Peers, Pirates, and Persuasion: Rhetoric in the Peer-to-Peer Debates.* West Lafayette, IN: Parlor Press, 2006.

Loughlan, Patricia. "Pirates, Parasites, Reapers, Sowers, Fruits, Foxes ... The Metaphors of Intellectual Property." *Sydney Law Review* 28 (2) (2006): 211–226.

Loughlan, Patricia. "'You Wouldn't Steal a Car': Intellectual Property and the Language of Theft." *European Intellectual Property Review* 29 (10) (2007): 401–405.

Madison, Michael J. "Rewriting Fair Use and the Future of Copyright Reform." *Cardozo Arts & Entertainment Law Journal* 23 (2) (2005): 391–418.

Marques, Jeannine. "Fair Use in the 21st Century: Bill Graham and *Blanch v. Koons.*" *Berkeley Technology Law Journal* 22 (1) (2007): 331–354.

Marshall, Lee. "Metallica and Morality: The Rhetorical Battleground of the Napster Wars." *Entertainment Law* 1 (1) (2002): 1–19.

McLeod, Kembrew. *Freedom of Expression: Overzealous Copyright Bozos and Other Enemies of Creativity.* New York: Doubleday, 2005.

McLeod, Kembrew, and Peter DiCola. *Creative License: The Law and Culture of Digital Sampling.* Durham, NC: Duke University Press, 2011.

Meese, James. "'It Belongs to the Internet': Animal Images, Attribution Norms and the Politics of Amateur Media Production." *M/C Journal* 17 (2) (2014). Available at http://journal.media-culture.org.au/index.php/mcjournal/article/view/782.

Meese, James. "Networked Subjects: Exploring the Author, User and Pirate through a Relationalist Lens." PhD thesis, Swinburne University of Technology, 2014.

Meese, James. "The Pirate Imaginary and the Potential of the Authorial Pirate." In *Piracy: Leakages from Modernity,* edited by Martin Fredriksson and James Avanitakis, 19–37. Sacramento, CA: Litwin Books, 2014.

Meese, James. "Resistance or Negotiation: An Australian Perspective on Copyright Law's Cultural Agenda." *Computers and Composition* 27 (3) (2010): 167–178.

Meese, James. "User Production and Law Reform: A Socio-Legal Critique of User Creativity." *Media, Culture & Society* 37 (5) (2015): 753–767.

Meese, James, Rowan Wilken, Bjorn Nansen, and Michael Arnold. "Entering the Graveyard Shift: Disassembling the Australian TiVo." *Television & New Media* 16 (2) (2015): 165–179.

Morreale, Joanne. "From Homemade to Store Bought: Annoying Orange and the Professionalization of YouTube." *Journal of Consumer Culture* 14 (1) (2014): 113–128.

Murray, Laura J., S. Tina Piper, and Kirsty Robertson. *Putting Intellectual Property in Its Place: Rights Discourses, Creative Labor, and the Everyday*. Oxford, UK: Oxford University Press, 2014.

Nedelsky, Jennifer. "Citizenship and Relational Feminism." In *Canadian Political Philosophy: Contemporary Reflections*, edited by Ronald Beiner and Wayne Norman, 131–146. New York: Oxford University Press, 2001.

Nedelsky, Jennifer. *Law's Relations: A Relational Theory of Self, Autonomy and Law*. Oxford, UK: Oxford University Press, 2011.

Nedelsky, Jennifer. "Reconceiving Rights as Relationship." *Review of Constitutional Studies/Revue d'etudes constitutionnelles* 1 (1) (1993): 1–26.

Negus, Keith. *Producing Pop*. London: Edward Arnold, 1992.

Ng, Alina. "When Users Are Authors: Authorship in the Age of Digital Media." *Vanderbilt Journal of Entertainment and Technology Law* 12 (4) (2010): 853–888.

Nowak, Raphaël. *Consuming Music in the Digital Age: Technologies, Roles and Everyday Life*. London: Palgrave Macmillan, 2016.

O'Brien, Damien, and Brian Fitzgerald. "Digital Copyright Law in a YouTube World." *Internet Law Bulletin* 9 (6 & 7) (2006): 71–74.

O'Brien, Damien, and Brian Fitzgerald. "Mashups, Remixes and Copyright Law." *Internet Law Bulletin* 9 (2) (2006): 17–19.

OECD Directorate for Science, Technology and Industry: Committee for Information, Computer and Communications Policy. *Participative Web: User-Created Content*. OECD, April 12, 2007. Available at https://www.oecd.org/sti/38393115.pdf

Okpaluba, Johnson. "Digital Sampling and Music Industry Practices, Re-Spun." In *Law and Creativity in the Age of the Entertainment Franchise*, edited by Kathy Bowrey and Michael Handler, 75–102. Cambridge, UK: Cambridge University Press, 2014.

Oliar, Dotan, and Christopher Sprigman. "There's No Free Laugh (Anymore): The Emergence of Intellectual Property Norms and the Transformation of Stand-Up Comedy." *Virginia Law Review* 94 (8) (2008): 1787–1867.

Pappalardo, Kylie. "A Tort Law Framework for Copyright Authorisation." PhD thesis, Australian Catholic University, 2015.

Patry, William. *How to Fix Copyright*. Oxford, UK: Oxford University Press, 2011.

Patry, William. *Moral Panics and the Copyright Wars*. Oxford, UK: Oxford University Press, 2009.

Patterson, Lyman Ray. *Copyright in Historical Perspective*. Nashville: Vanderbilt University Press, 1968.

Patterson, Lyman Ray. "Folsom v. Marsh and Its Legacy." *Journal of Intellectual Property Law* 5 (Spring 1998): 431–452.

Patterson, Lyman Ray, and Stanley W. Lindberg. *The Nature of Copyright: A Law of Users' Rights*. Athens: University of Georgia Press, 1991.

Paulu, Burton. *Television and Radio in the United Kingdom*. Minneapolis: University of Minnesota Press, 1981.

Philip, Kavita. "What Is a Technological Author? The Pirate Function and Intellectual Property." *Postcolonial Studies* 8 (2) (2005): 199–218.

Potts, Jason, John Hartley, John Banks, Jean Burgess, Rachel Cobcroft, Stuart Cunningham, and Lucy Montgomery. "Consumer Co-Creation and Situated Creativity." *Industry and Innovation* 15 (5) (2008): 459–474.

Raustiala, Kal, and Christopher Sprigman. *The Knockoff Economy: How Imitation Sparks Innovation*. Oxford, UK: Oxford University Press, 2012.

Raustiala, Kal, and Christopher Sprigman. "The Piracy Paradox: Innovation and Intellectual Property in Fashion Design." *Virginia Law Review* 92 (8) (2006): 1687–1777.

Reyman, Jessica. *The Rhetoric of Intellectual Property*. New York: Routledge, 2010.

Rimmer, Matthew. *Digital Copyright and the Consumer Revolution: Hands Off My iPod*. Cheltenham, UK: Edward Elgar Publishing, 2007.

Rimmer, Matthew. "An Elegy for Greg Ham: Copyright Law, the Kookaburra Case, and Remix Culture." *Deakin Law Review* 17 (2) (2013): 385–423.

Rimmer, Matthew. "Hail to the Thief: A Tribute to Kazaa." *University of Ottawa Law and Technology Journal* 2 (1) (2005): 173–218.

Rimmer, Matthew. "Robbery under Arms: Copyright Law and the Australia-United States Free Trade Agreement." *First Monday* 11 (3) (2006). Available at firstmonday .org/ojs/index.php/fm/article/view/1316.

Rimmer, Matthew. "Trick or Treaty?: The Australian Debate over the Anti-Counterfeiting Trade Agreement." In *The ACTA and the Plurilateral Enforcement Agenda: Genesis and Aftermath*, edited by Pedro Roffe and Xavier Seuba, 169–201. Cambridge, UK: Cambridge University Press, 2015.

Rose, Mark. *Authors and Owners: The Invention of Copyright*. Cambridge, MA: Harvard University Press, 1993.

Rose, Mark. "Nine-Tenths of the Law: The English Copyright Debates and the Rhetoric of the Public Domain." *Law and Contemporary Problems* 66 (1/2) (2003): 75–87.

Rugg, Adam, and Ben Burroughs. "Periscope, Live Streaming and Mobile Video Culture." In *Geoblocking and Global Video Culture*, edited by Ramon Lobato and James Meese, 64–73. Amsterdam: Institute of Network Cultures, 2016.

Sag, Matthew. "The Pre-History of Fair Use." *Brooklyn Law Review* 76 (4) (2011): 1371–1412.

Samuelson, Pamela. "Why Copyright Law Excludes Systems and Processes from the Scope of Its Protection." *Texas Law Review* 85 (7) (2007): 1921–1977.

Scassa, Teresa. "Acknowledging Copyright's Illegitimate Offspring: User-Generated Content and Canadian Copyright Law." In *The Copyright Pentalogy: How the Supreme Court of Canada Shook the Foundations of Canadian Copyright Law*, edited by Michael Geist, 431–454. Ottawa, CA: University of Ottawa Press, 2013.

Schietinger, John. "Bridgeport Music, Inc. v. Dimension Films: How the Sixth Circuit Missed a Beat on Digital Music Sampling." *DePaul Law Review* 55 (1) (2005): 209–248.

The Senate of Australia. *Select Committee on the Free Trade Agreement between Australia and the United States of America: Final Report*. Canberra: Parliament House, 2004.

Seville, Catherine. *Literary Copyright Reform in Early Victorian England: The Framing of the 1842 Copyright Act*. Cambridge, UK: Cambridge University Press, 1999.

Sherman, Brad, and Lionel Bently. *The Making of Modern Intellectual Property Law*. Cambridge, UK: Cambridge University Press, 1999.

Shi, Xiao Xiang. "Towards a Relational Theory of Copyright Law: Reconfiguring Author's Economic Rights to Facilitate Knowledge Growth in Networked Information Societies." PhD thesis, Queensland University of Technology, 2010.

Shi, Xiao Xiang, and Brian Fitzgerald. "A Relational Theory of Authorship." In *Knowledge Policy for the 21st Century: A Legal Perspective*, edited by Brian Fitzgerald and Mark Perry. Toronto: Irwin Law, 2011. Available at https://www.irwinlaw.com/content_commons/knowledge_policy_for_the_21st_century.

Shirky, Clay. *Here Comes Everybody: The Power of Organizing without Organizations*. New York: Penguin, 2008.

Simpson, Alfred William Brian. *Legal Theory and Legal History*. London: The Hambledon Press, 1987.

Sirluck, Ernest. "Areopagitica and a Forgotten Licensing Controversy." *Review of English Studies* 11 (43) (1960): 260–274.

Siskind, Shira. "Crossing the Fair Use Line: The Demise and Revival of the Harry Potter Lexicon and Its Implications for the Fair Use Doctrine in the Real World and on the Internet." *Cardozo Arts & Entertainment Law Journal* 27 (1) (2009): 291–311.

Smith, Paul. *Discerning the Subject*. Minneapolis: University of Minnesota Press, 1988.

Snickars, Pelle, and Patrick Vonderau, eds. *The YouTube Reader*. Stockholm: National Library of Sweden, 2009.

St. Clair, William. *The Reading Nation in the Romantic Period*. Cambridge, UK: Cambridge University Press, 2004.

Stevens, Jessica. "The Secondary Sale, Copyright Conundrum—Why we Need a Secondary Market for Digital Content." *Australian Intellectual Property Journal* 26 (4) (2016): 179–194.

Streeter, Thomas. 2010. *The Net Effect: Romanticism, Capitalism and the Internet*. New York: New York University Press.

Sundaram, Ravi. *Pirate Modernity: Delhi's Media Urbanism*. New York: Routledge, 2010.

Sunder, Madhavi. *From Goods to a Good Life: Intellectual Property and Global Justice*. New Haven, CT: Yale University Press, 2012.

Tang, Xi Yin. "The Artist as Brand: Toward a Trademark Conception of Moral Rights." *Yale Law Journal* 122 (1) (2012): 218–257.

Thomas, Julian. "The Old New Television and the New: Digital Transitions at Home." *Media International Australia* 129 (2008): 91–103.

TorrentFreak. *The Speculative Invoicing Handbook*. Self-Published, 2009.

Towse, Ruth. "The Quest for Evidence on the Economic Effects of Copyright Law." *Cambridge Journal of Economics* 37 (5) (2013): 1187–1202.

Turner Hopkins. *Report for Ofcom: The Value of User-Generated Content*. Warrington, UK: Ofcom, 2013.

Tushnet, Rebecca. "Copy This Essay: How Fair Use Doctrine Harms Free Speech and How Copying Serves It." *Yale Law Journal* 114 (2004): 535–590.

Tushnet, Rebecca. "Worth a Thousand Words: The Images of Copyright." *Harvard Law Review* 125 (3) (2011): 683–759.

Urban, Jennifer, Joe Karaganis, and Brianna Schofield. *Notice and Takedown in Everyday Practice*. Berkley, CA: Berkley Law, 2016.

Vaidhyanathan, Siva. *Copyrights and Copywrongs: The Rise of Intellectual Property and How It Threatens Creativity*. New York: New York University Press, 2003.

Velthuis, Olav. "Symbolic Meanings of Prices: Constructing the Value of Contemporary Art in Amsterdam and New York Galleries." *Theory and Society* 32 (2) (2003): 181–215.

Vertigan, Phoebe. "EMI Songs Australia Pty Ltd v Larrikin Music Publishing Pty Ltd (2011) 191 FCR 444." *University of Tasmania Law Review* 30 (2) (2011): 155–163.

Vonderau, Patrick. "The Video Bubble: Multichannel Networks and the Transformation of YouTube." *Convergence: The International Journal of Research into New Media Technologies* 22 (4) (2016): 361–375.

Walker, Jesse. *Rebels on the Air: An Alternative History of Radio in America.* New York: New York University Press, 2001.

Weatherall, Kimberlee. "Of Copyright Bureaucracies and Incoherence: Stepping Back from Australia's Recent Copyright Reforms." *Melbourne University Law Review* 31 (3) (2007): 967–1174.

Wilson, Stephen. "Rewarding Creativity: Transformative Use in the Jazz Idiom." *Pittsburgh Journal of Technology Law and Policy* 4 (2003–2004): 1–32.

Woodmansee, Martha. *The Author, Art, and the Market.* New York: Columbia University Press, 1994.

Woodmansee, Martha. "The Genius and the Copyright: Economic and Legal Conditions of the Emergence of the 'Author.'" *Eighteenth-Century Studies* 17 (4) (1984): 425–448.

Woodmansee, Martha, and Peter Jaszi. *The Construction of Authorship: Textual Appropriation in Law and Literature.* Durham, NC: Duke University Press, 1994.

Woolgar, Steve. "Configuring the User: The Case of Usability Trials." *Sociological Review* 38 (S1) (1990): 58–99.

Yankwich, Leon. "What Is Fair Use?" *University of Chicago Law Review* 22 (1) (Autumn 1954): 203–215.

Yu, Peter. "Can the Canadian UGC Exception Be Transplanted Abroad?" *Intellectual Property Journal* 26 (2) (2014): 175–203.

Yudkin, Jeremy. "Beethoven's 'Mozart' Quartet." *Journal of the American Musicological Society* 45 (1) (1992): 30–74.

Index

Italic page numbers refer to figures.